Winter Twigs of Arkansas

A Field Guide to Deciduous Woody Plants

G. Thomas Clark
Professor of Biology
Hendrix College

Illustrations by
T. Patterson Clark

Rose Publishing Company
Little Rock, Arkansas

To my wife
Natalie Williams Clark

Library of Congress Catalog Card Number 81-50399
ISBN No. 0-914546-35-X

A tree does not die in winter;
nor is it resurrected in spring.
It waits. — **Hesse.**

Acknowledgments

The original drawings are the work of T. Patterson Clark. I gratefully acknowledge his dedication to details and quality.

This work has been funded by an unrestricted Faculty Grant from the Lilly Endowment of 1975-77, the Hendrix College Faculty Grants Program, and the Biology Department. I have received constant encouragement from Dr. Arthur A. Johnson along with generous support of materials from the department.

Several classes of botany students have been directly involved in producing this field guide. Their cautious guidance and patient criticism is deeply appreciated.

My thanks go to Nancy Ross for the cover design and to Holly Clark for the original farmstead photograph. Jeannie Wilder is to be praised for the tedium she endured with the map graphics.

Many individuals have cooperated in my search for twigs and I recognize the special efforts made by David M. Johnson and L. P. Lowman.

I wish to thank Dr. Bland Crowder and Holly Clark for their critical reading of the manuscript.

Contents

G. THOMAS CLARK is The Louis and Charlotte Cabe Distinguished Professor of Biology at Hendrix College in Conway, Arkansas. He teaches plant science courses in the liberal arts curriculum. He directs the herbarium, supervises the college greenhouse, and serves actively in the management of campus landscaping.

Dr. Clark holds the Bachelor of Science degree in education from Eastern Illinois University. He received the Master of Science degree in botany from the University of Michigan and the Doctor of Philosophy degree in plant ecology from the University of Arkansas, Fayetteville.

His publications include **Winter Twigs, Forest Communities of Crowley's Ridge,** and **A Preliminary Ecological Study of Crowley's Ridge.** His continuing research deals with plant community composition and distribution.

He is a member of the American Association for the Advancement of Science, The American Institute of Biological Sciences, The Ecological Society of America, The Botanical Society of America, The Scientific Research Society of America, The Arkansas Academy of Science, and The Horticulture Society of America.

Preface

To learn that buds are on woody plants at the time of leaf fall is a revelation that runs contrary to the common belief that trees, shrubs, and vines "bud out" in the spring. If this has been your experience, then you will understand my initial interest in the winter aspects of deciduous woody plants.

This field guide is the result of a personal search for more information that has spilled over into the classroom and academic study. It is directed to the public, to students of all ages, to naturalists, and to scientists. The descriptive terminology is limited to the words that give the greatest accuracy and information. The illustrated glossary gives definition to those words which generally may not be understood. Easy access to the information was deemed important because there are very practical reasons for making a quick decision in the winter fields and forests. A continuing use of the field guide can lead to an ability to recognize species immediately by form, structure, and habitat. The ultimate reward of the organized study is an aesthetic appreciation and understanding which surpasses mere identification of individual plants.

This second edition follows the definitive work of Dr. Edwin B. Smith, *An Atlas and Annotated List of Vascular Plants of Arkansas,* 1978, which updates the nomenclature and knowledge of Arkansas flora. Interpretation of species and habitats also relies on the unpublished work of Dr. Gary Tucker, *A Guide To The Woody Flora of Arkansas,* 1975. Both are recommended for complete listings and definitive explanations of Arkansas flora and Arkansas woody flora, respectively. Distribution maps were derived from these two sources and from my own field experiences.

A definitive list of deciduous woody plants is included in order to show the scope of this work and to provide a checklist for the reader. Many species were eliminated by necessity. In the winter condition, it is improbable that species of some genera can be confidently separated because of the limited amount of structural information. Moreover, the semi-woody condition of some species disallows for a clear structural analysis.

G. Thomas Clark

Conway, Arkansas
July 22, 1980

Introduction

The winter features of woody plants show structural details sufficient for positive identification of a species. The combined information about the specific characteristics of the twig, the general growth form, and a recognition of habitats, provides a wealth of pertinent evidence to determine its identity.

Twigs. A twig is the branch tip of a woody plant. Usually lacking lateral branches but including several nodes and buds, a twig represents the growth and structure of one growing season. It commonly changes color beyond its starting point. The season's growth can amount to several inches or several feet. Growth in diameter will also vary with the species.

Buds. Buds are the most conspicuous features of twigs. Their arrangement may be alternate, opposite, subopposite, or whorled. They project from the surface of the twig, may be stalked, or embedded. Terminal buds or pseudo-terminal buds are accompanied by lateral buds along the sides of the twigs at nodes. Lateral buds commonly appear singly but can be accompanied by accessory buds that are superposed or collateral in position. Buds are classified according to their content: vegetative, flower, or mixed. Buds may be enclosed by imbricate, valvate, or conicate bud scales. Those lacking scales are regarded as naked buds.

Scars. In the spring, with the expansion of buds, the scales are usually shed, leaving a band of scars around the twig. Thus, terminal bud-scale scars mark the beginning of each year's growth and provide a means of estimating the age of young branches. Bud scales are persistent in some species and appear at the base of branches.

Other scars are also definitive features of twigs. Leaf scars remain as prominent, conspicuous features on twigs and branches long after the leaves have fallen. They show a great variety of form among the common trees, shrubs, and vines. One or more vascular bundle scars appear as dots or lines in each leaf scar. Patterns of scars, representing the ends of single bundles or groups of bundles, are distinctive for a species. Stipular scars appear as lateral continuations of the leaf scar in some species and may encircle the twig. The loss of any other lateral extension of the twig will leave characteristic scars, viz., terminal bud scars, lateral bud scars, fruit scars, branch scars.

Surfaces. Thorns, spines, prickles, bristles, ridges, corky wings, and lenticels are common projections and outgrowths on twigs. Hairiness, pubescence, waxiness, or absence of these, are surface characteristics that aid in differentiating woody plants.

Bark. The bark of every woody plant has its special texture and color that change with age, vitality, and environment. The old bark turns to cork

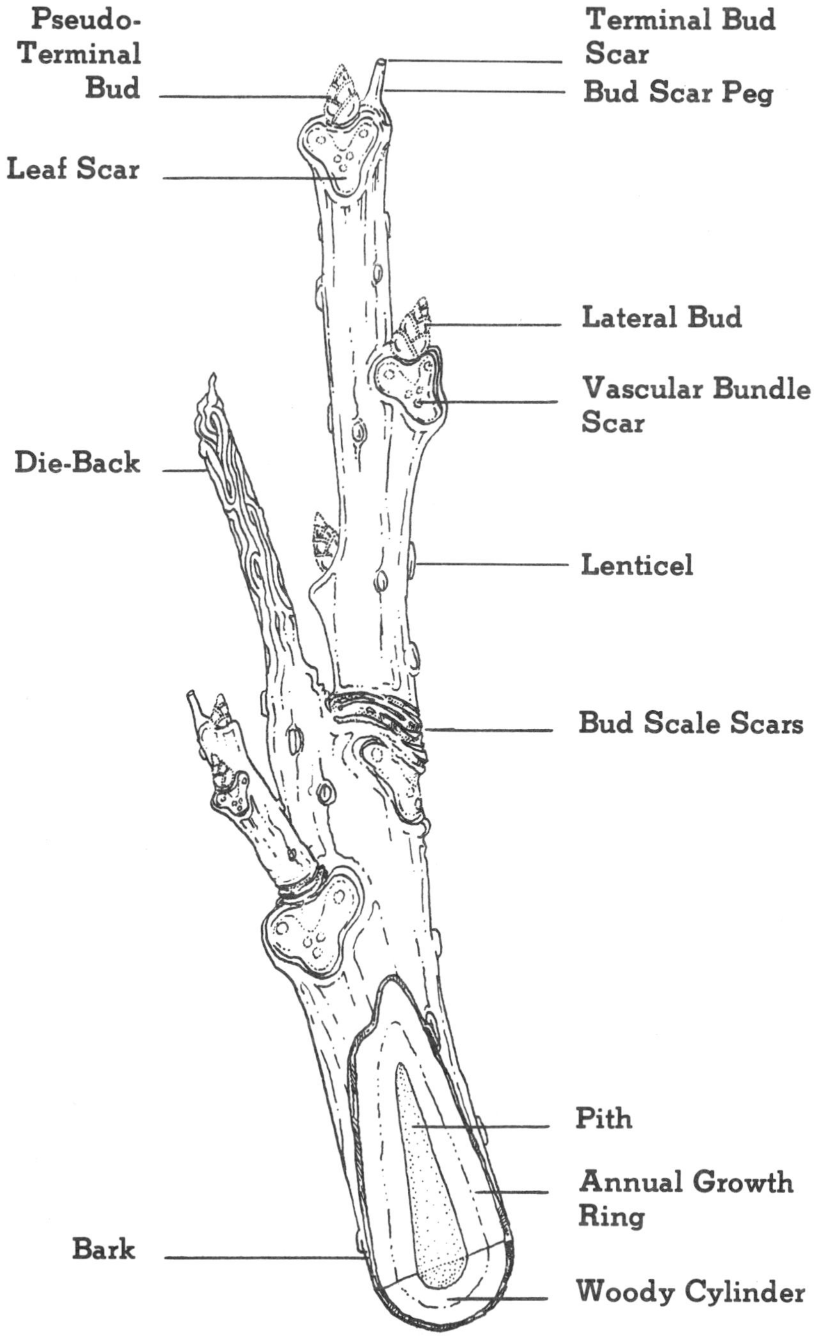

Features of Woody Stems

and splits as the new bark grows beneath it. Bark characteristics help significantly in making specific identification.

Growth form. The general appearance of deciduous plants in winter is determined largely by the way it grows. The disposition of buds on twigs determines the growth form. The number, form, and size of trunks; the texture and shape of the crown; and the number, size, and direction of the growth of the twigs, are all used to form a visual impression of a species. The natural tendency toward a uniform growth pattern is altered by habitat; therefore, variability is to be expected.

Habitat. A recognition of different habitats is an aid in plant identification. Habitat recognition provides information about exposure to the sun, predictable temperature extremes, soil capabilities, and availability of water. The response of plants to these variable factors determines their success in a habitat. Woody plants grow with degrees of success over a range of habitats. This range is wide for some, narrow for others. Distribution of a species indicates its tolerance to habitat conditions. Site preference can often be useful in identification where species occur naturally. The importance of site preference is diminished by transplantation and cultivation.

Plant Communities. A plant community is an assemblage of species populations in a particular area. Recurrence of these plant aggregates shows that there is a high probability that certain species will occur together. The plant communities are named according to the dominant species, viz., Sugarberry-Elm-Ash, Oak-Hickory-Pine, etc.

Vegetation Types. Broader conglomerates of plant communities amount to vegetation types. The general appearance of the life- or growth-form of the collective plant communities of an area becomes the significant parameter. Variation in appearance is the first justification for classification and naming the vegetation types of Arkansas (Pell, 1980):

Forest Vegetation: Dominated by trees.

Glade/Outcrop Vegetation: Trees and vegetation locally sparse and reduced by bedrock outcroppings.

Herbaceous Vegetation: Dominated by herbs and grasses.

Savanna Vegetation: Dominated by grasslands with sparse trees.

Scrub/Shrub Vegetation: Dominated by low scrub or shrubs of wetlands or mountains.

Natural Environment. The abrupt changes in elevation from lowlands to highlands in Arkansas are detectable even to the untrained eye. These sharp contrasts in the topography are the result of several cycles of geological events at long time intervals. Two different kinds of mountain building processes, marine sedimentation, scouring and sedimentation by major streams, vast erosion of uplands, and deposition of wind-blown soils, account for the major land features.

In the uplands, the Ozark Mountains are characterized by plateaus, rimrocks, complex stream patterns, and horizontal rock layers in roadcuts and cliffs. The Ouachita Mountains have a different appearance. They run east and west, forming long valleys and rugged peaks of folded rock layers. The collision of the Ouachitas with the Ozarks resulted in a downwarping through which the Arkansas River now flows and cuts its valley.

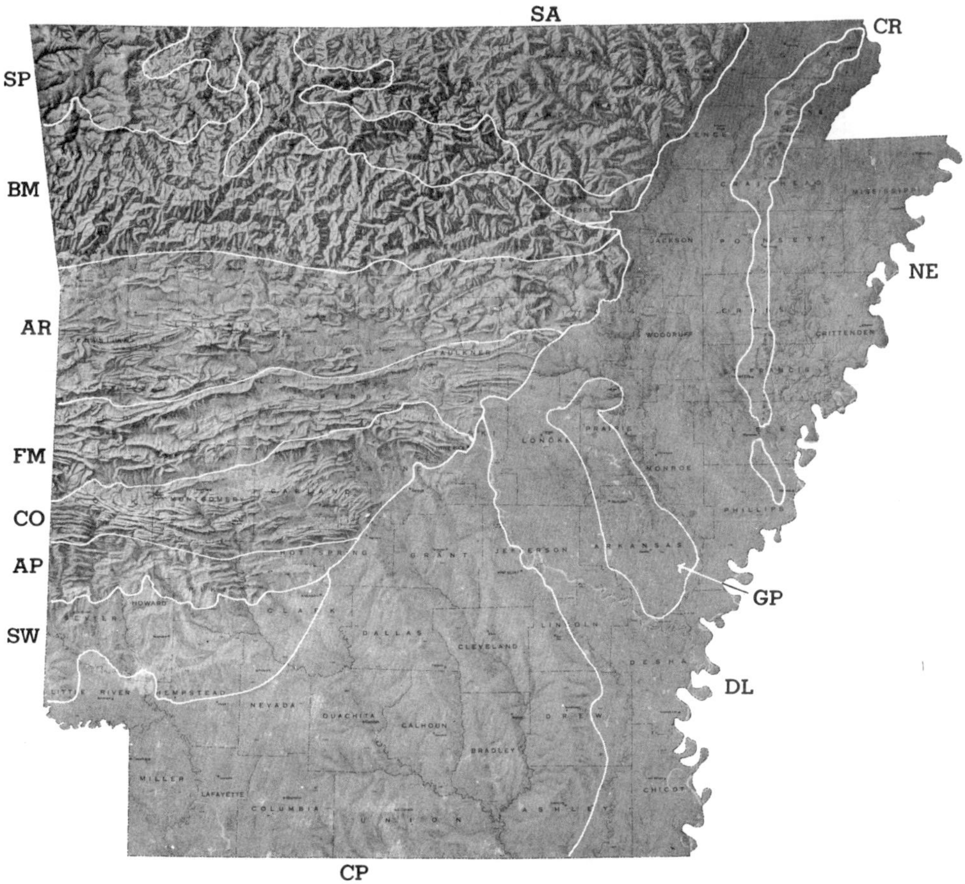

INTERIOR HIGHLANDS

Ozark Mountains

SA - Salem Plateau
SP - Springfield Plateau
BM - Boston Mountains

Ouachita Mountains

AR - Arkansas River Valley
FM - Fourche Mountains
CO - Central Ouachita Mountains
AP - Athens Piedmont Plateau

LOWLANDS

West Gulf Coastal Plain

CP - Coastal Plain
SW - Southwestern Arkansas

Crowley's Ridge - CR

Mississippi Alluvial Plain

DL - Delta
GP - Grand Prairie
NE - Northeastern Arkansas

The Gulf Coastal Region, the lowlands, was formerly covered by shallow embayments of the Gulf of Mexico. Some of these old sediments, in the area known as the Mississippi Alluvial Plains and Terraces, were scoured out by melt waters of the last glacier and replaced by other sediments. However, the old marine sediments are still evident as the West Gulf Coastal Plains and as the anomalous Crowley's Ridge. Both areas were subsequently eroded dramatically by streams, and Crowley's Ridge was covered by a thick mantle of loess, a fine-textured windblown soil.

These natural features are systematically classified (Foti, 1974) according to Natural Divisions (Map) and the system is a practical fieldguide to habitat recognition.

How to Use the Guide

- **Use a 10X or 15X hand lens to examine twigs.**
  Magnification is necessary in order to verify the structural details of all twigs.

- **Observe living twigs.**
  Living twigs have plump, vigorous buds; a full, tight bark with no shrivels; intense coloration; and a moist, green, inner-bark.

- **Examine several twigs on the same plant in order to determine normal characteristics.**
  The color, texture, arrangement and covering of buds, shape of the leaf scar, condition of the terminal bud, presence of thorns, spines, and prickles, odor, and taste, are crucial to correct identification.

- **Classify the plant and twig by the terms and symbols on the opposite page.**
  The terms are defined in the glossary and each has a thumb tab symbol.

- **Use the thumb tabs to find and enter the appropriate subdivisions.**
  The major divisions—Trees/Shrubs and Vines—are subdivided once, twice, or thrice. Subdivisions with numerous species are summarized on a page with schematics and details that isolate small groups of similar species. Key characteristics of these are listed and lead to a tentative identification. A page number reference follows.

- **Study the illustrations, descriptions, and maps to identify the specimen at hand.**
  Structural details of the twig, combined with fruit-structure information, geographical location, and habitat recognition, should provide enough information for positive identification.

- **Check the list in the back of the book for the plant-family relationships.**
  The appropriate family classification is given at the end of each species description. An alphabetized reference list for all deciduous, woody plants of Arkansas is appendixed.

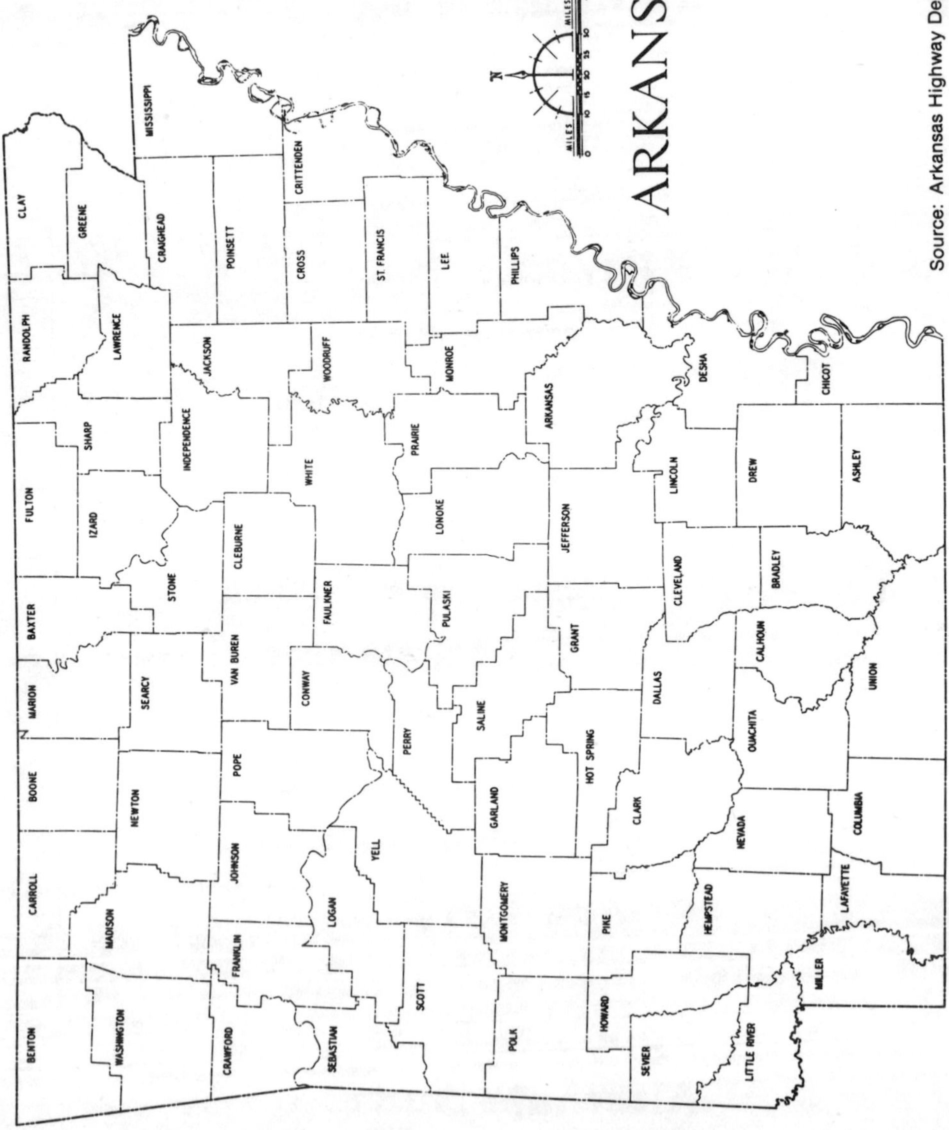

Source: Arkansas Highway Department

TREES/SHRUBS — Adventitious Buds

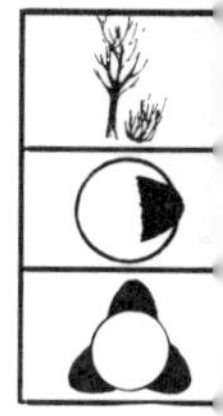

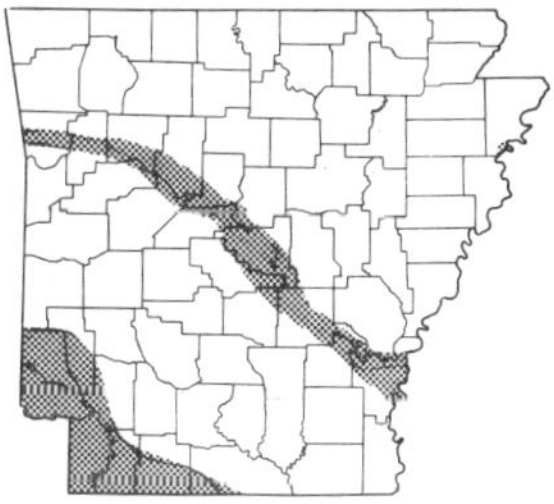

TAMARISK — *Tamarix chinensis.* Twigs long-tapering, terete, bright red-brown, and raspy due to raised branch scars. There are no leaf scars. Lateral buds arise adventitiously from the center of the branch scars. Tip die-back is common leading to irregular and widely spread branching. Introduced from China for ornamentals, windbreaks, and sand binding. A shrubby, small tree that forms thickets on sandy beaches of major rivers. *Tamaricaceae.*

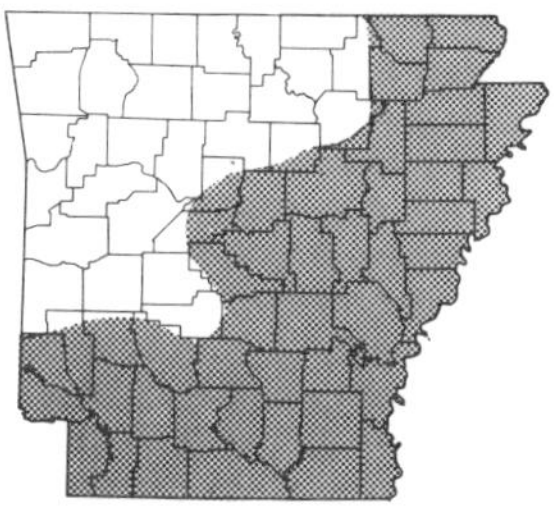

BALD CYPRESS — *Taxodium distichum.* Terminal bud present. Leaf scars lacking; only branch scars are present. Lateral buds adventitious and randomly alternate. Twigs smooth, shiny, and bright brown. Bark of older branches thin, finely shredded, and bright brown. Young trees have strong, excurrent growth form with finely diffuse lateral branches. Crowns of old trees become flattened and irregular. Dominant trees of swamps and shallow lakes. Occurs in pure stands or with sour gum, buttonbush, and overcup oak. *Taxodiaceae.*

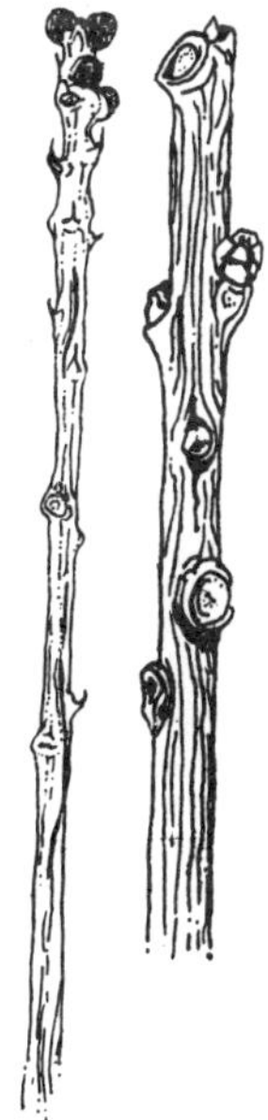

TREES/SHRUBS — Whorled Buds

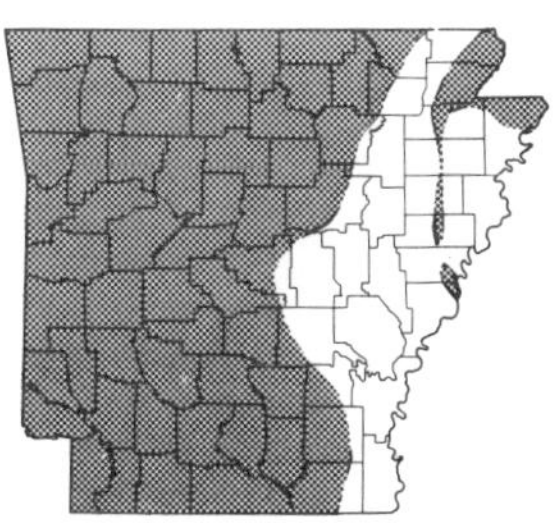

CATALPA — *Catalpa speciosa.* Twig tips reduced in size, often withered. Buds and leaf scars whorled. Leaf scars are elliptical with an inner circle of vascular bundle scars. Growth of two of the three lateral buds at the terminal node produces the forked branching pattern of the tree. Long, cylindrical fruits containing shaggy-winged seeds persist into winter. *Bignoniaceae.*

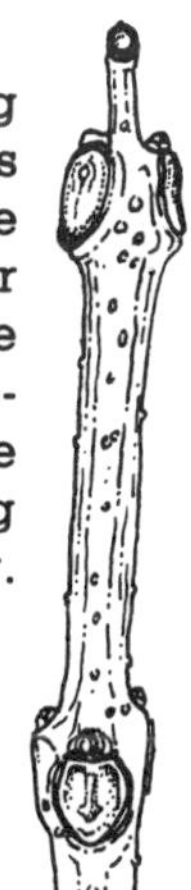

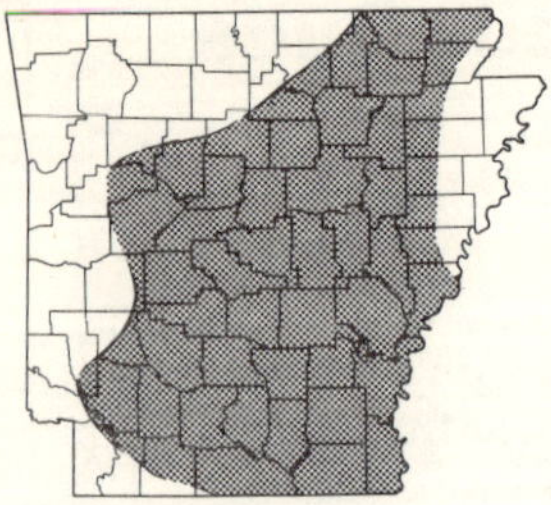

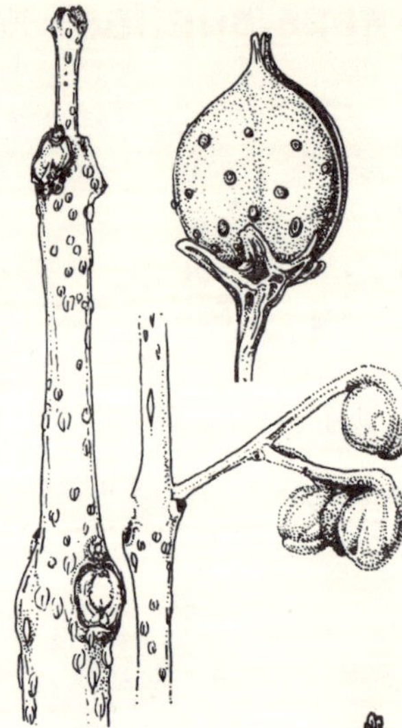

PRINCESS TREE — *Paulownia tomentosa.* A Chinese tree propagated widely by sprouts, seeds, greenwood cuttings, and by new unfolding leaves. Conspicuous in winter by old, two-parted, rattlebox-fruits, and by large terminal clusters of dull, gold, tomentose, beaked flower buds. Flower stalks arise from lateral buds and support five flower buds. Vegetative twigs show die-back at the tips. Leaf scars and lateral buds opposite, occasionally whorled, on glabrous, olive drab twigs. Lenticels white, numerous, and warty. Leaf scars round to elliptical with a central circle of vascular bundle scars. Bud scales short and fleshy. Pith hollow or intermittent. *Scrophulariaceae.*

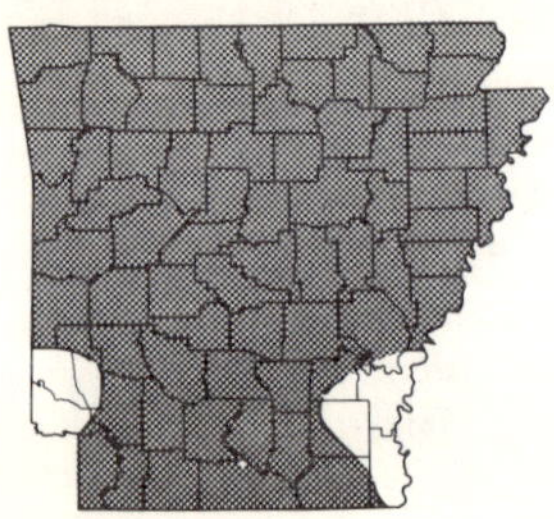

BUTTONBUSH — *Cephalanthus occidentalis.* The terminus of twigs shows die-back. Lateral buds whorled or opposite. Leaf scars are elliptical with persistent black stipules that nearly connect the leaf scars. Vascular bundle scars form a crescent pattern in the central leaf scar. Bark dark gray to brown. A thicket forming shrub of wet-logged soils and lake margins. *Rubiaceae.*

TREES/SHRUBS Naked Opposite Buds

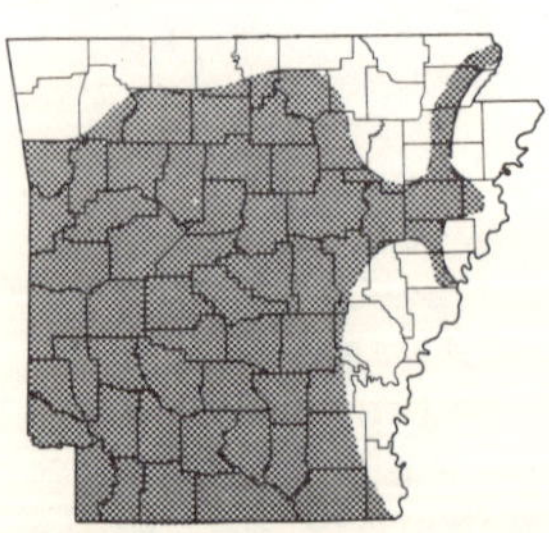

FRENCH MULBERRY — *Callicarpa americana.* Terminal bud naked, gray, dull, and covered with star-shaped hairs. Twig tip is scurfy and appears dull; covered with star-shaped hairs and resinous dots. Rubbing the resin produces a sharp resin odor. Lateral buds naked, hairy, and opposite with superposed buds and fruit stalks. Leaf scars elliptical and notched at the top; depressed with a center crescent of vascular bundle scars. Stipule scars, wart-like at the sides of each leaf scar. Older twigs dull gray with scattered warty lenticels and resin dots. Clusters of red-purple, fleshy fruits surround the stem at several nodes in early fall; if not eaten by birds, falls away early, leaving wiry, branched fruit stalks. A handsome small shrub of dry woods. *Verbenaceae.*

SHRUBS

SMALL TREES

TREES

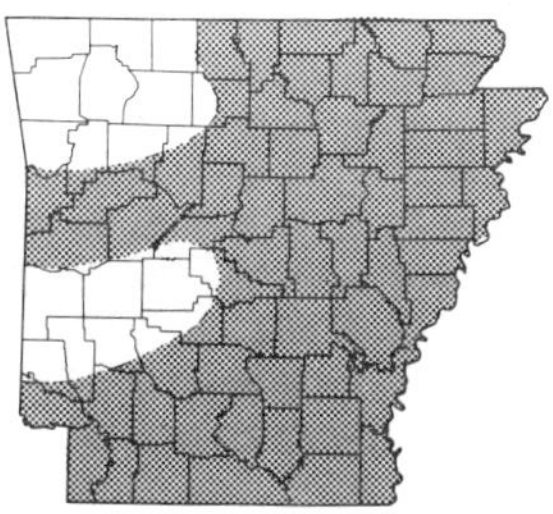

SWAMP PRIVET — *Forestiera acuminata.* A shrub or slender, multi-trunked tree of swamps and overflow lands. Noticed in the field by its dull olive branches which fade subtly to a buff brown. Light corky lenticels give a warty appearance. Terminal bud present on delicate hirsute twigs with opposite to subopposite lateral buds. All buds are imbricate and straw-colored. Leaf scars raised, half-round, and with straw-colored rims. Wetlands. *Oleaceae.*

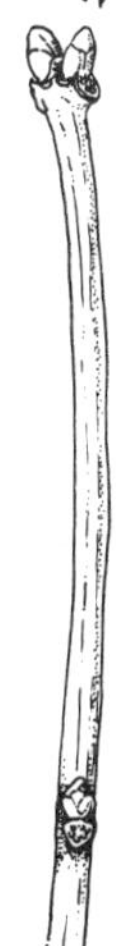

AMERICAN BLADDERNUT — *Staphylea trifolia.* A slender wide-topped shrub or small tree with smooth green-mottled twigs and greenish old stems striped with corky ridges. Twigs appear blunt because of a terminal bud scar flanked by opposite buds of equal size. Bud scales fleshy. Opposite leaf scars half-round, raised, attended by stipular scars, and showing six vascular bundle scars arranged in a circle. A strong odor is released from bruised bark. Bladdery, inflated, three-horned fruits appear in early winter. Found in borders of upland woods in moist sites. *Staphyleaceae.*

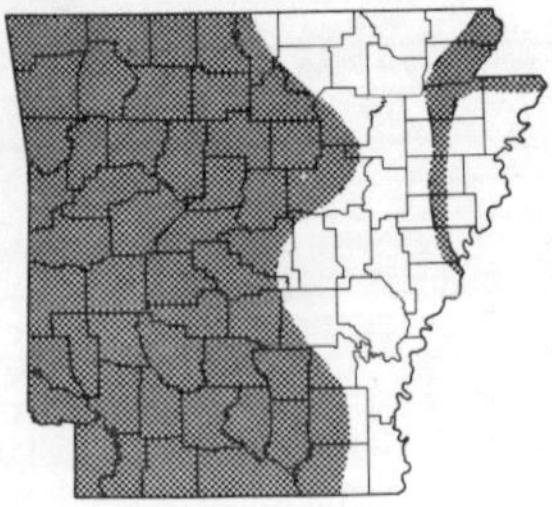

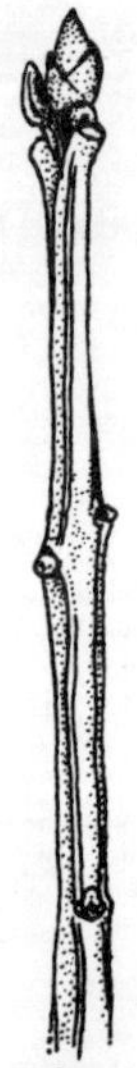

STRAWBERRY BUSH — *Euonymus americanus.* Terminal bud present. Lateral buds sub-opposite, imbricate, and sharp. Twigs green with four ridges. Conspicuous in early fall by pink fruits opening and showing orange, pulpy seeds. Seen in winter as a supple-limbed, bright-green shrub common to floodplains. *Celastraceae.*

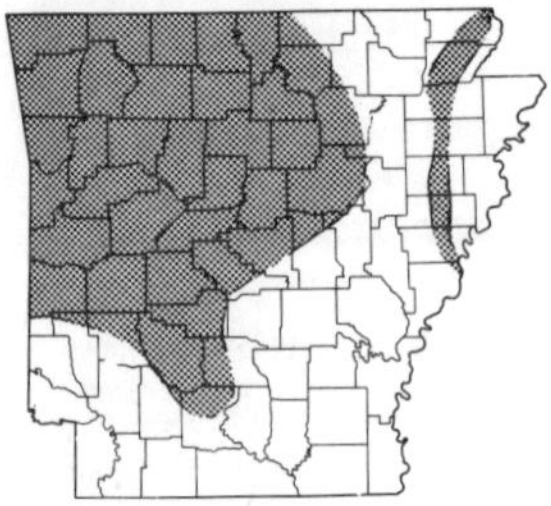

WILD HYDRANGEA — *Hydrangea arborescens.* Terminal bud present. Lateral buds opposite and imbricate. Twig tan and brittle. Pith large in relation to twig diameter. One can find persistent, dried, flower clusters with marginal sterile flowers. A formal-appearing, stiff, clumpy shrub of deep, moist, highland woods. *Saxifragaceae.*

COMMON ELDERBERRY — *Sambucus canadensis.* A brittle, pithy, scarcely woody, multi-stemmed shrub of rich soils and moist woods. Bark warty, reddish-brown, easily removed and rank smelling. The twig tips show die-back or terminal bud scars. Imbricated lateral opposite buds are superposed over broad, concave leaf scars. Leaf scars show five vascular bundle scars and are joined around the stem by stipular scars. Terminal fruit clusters, flat-topped with drooping sides, bear purple-black, juicy fruits with three seeds. Fruits and seeds are spread by birds and people who eat them. The white elderberry pith facilitates laboratory hand-sectioning of plant materials. Long de-pithed internodes of the larger stems make potential paperwad popguns and crude flutes. *Caprifoliaceae.*

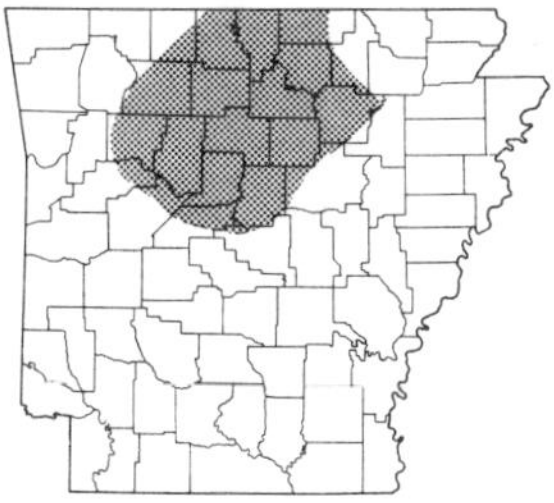

MOCK ORANGE — *Philadelphus hirsutus.* Low, loose arching shrubs with opposite perpendicular side branches. Twigs dull red-brown flaking away to a smooth, pale brown. Older limbs shredding into thin, netted patterns. Terminal bud with loose imbricate scales. Long matted hairs often make a webbing at the base of the bud. Lateral buds opposite and similar to the terminal bud. Leaf scars raised, crescent-shaped and joined around the twig. Vascular bundle scars in four groups. Long, sparse hairs cause hirsute twigs. Small, disjunctive populations of these shrubs are found on ledges, cliffs, and slopes of upland deciduous and shortleaf pine forests. *Saxifragaceae.*

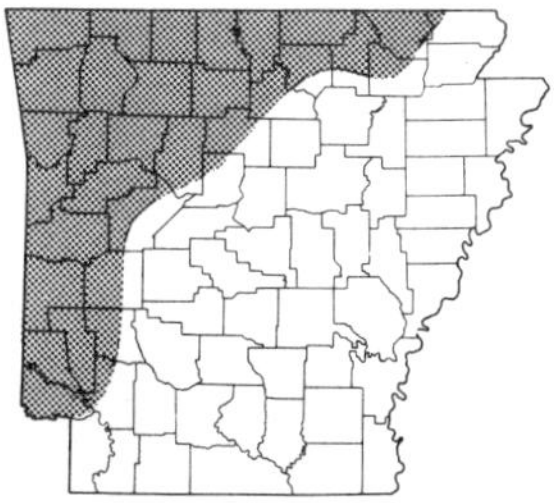

OHIO BUCKEYE — *Aesculus glabra.* Terminal bud brown, dry, imbricate and large compared to the lateral buds. Bud scales in four to five ranks from base to the tip; margins ciliate. Scale tips sharp pointed and extended. Buds opposite with broad, flat obtriangular leaf scars showing three patches of vascular bundle scars. Leathery husks of old fruit with coarse, soft spines. *Hippocastanaceae.*

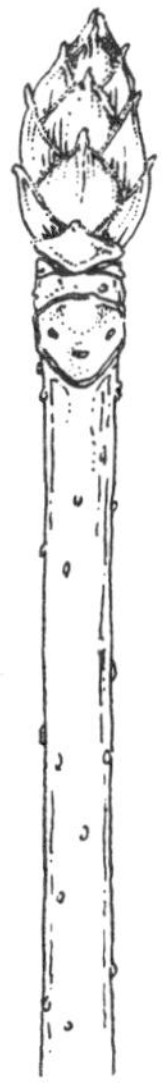

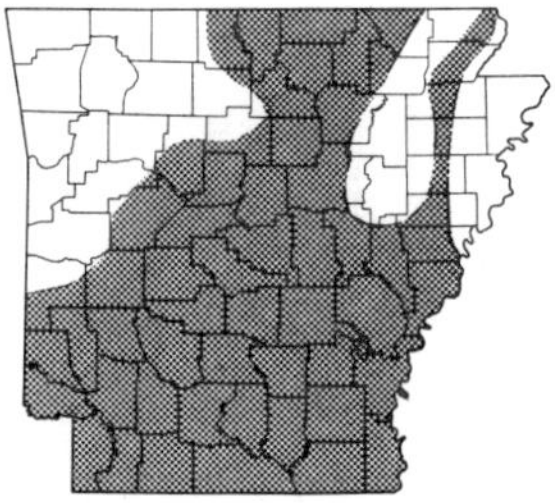

RED BUCKEYE — *Aesculus pavia.* Terminal bud brown, dry, imbricate, and large compared to the lateral buds. Bud scales in three to four ranks from base to tip; margins membraneous. Scale tips short and appressed. Buds opposite with broad, flat obtriangular leaf scars showing three patches of vascular bundle scars. Annual flowering produces forked branching. A small tree with a sparse, open crown. *Hippocastanaceae.*

SUGAR MAPLE — *Acer saccharum.* Terminal bud imbricate and sharp. Lateral buds opposite. Leaf scars chevron-shaped with three groups of vascular bundle scars. Twigs brown to gray, smooth and shiny. A large tree with a dense, broad crown of smooth, gray limbs. *Aceraceae.*

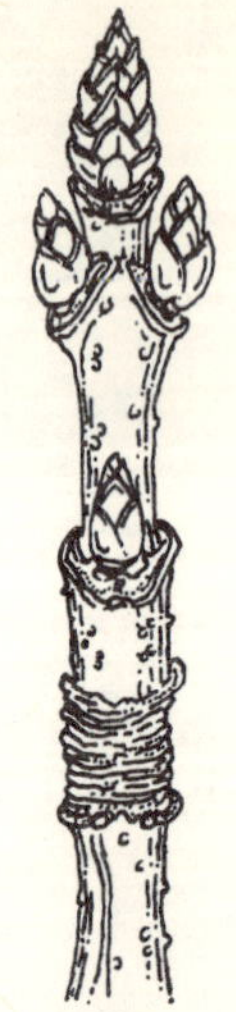

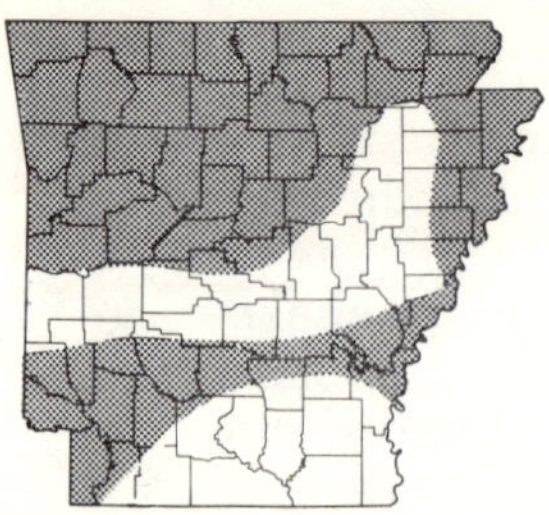

SILVER MAPLE — *Acer saccharinum.* Terminal bud blunt, imbricate, red, and shiny. Lateral buds opposite with chevron-shaped leaf scars. Vascular bundle scars in three groups. Globose collateral flower buds common. Twigs red to orange with a rank odor when bruised. Bud scale margins ciliate. Trees found where there is uniform soil moisture during the growing season. *Aceraceae.*

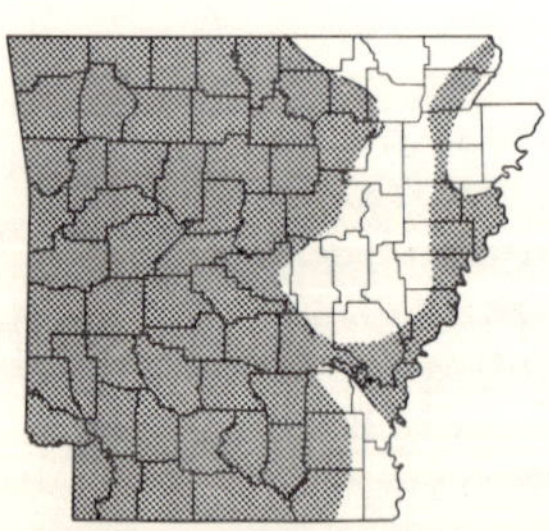

RED MAPLE — *Acer rubrum.* Terminal bud imbricate, glabrous, and dull red. Bud scales red with dense white tomentum. Lateral buds opposite with a narrow chevron-shaped leaf scar. Three vascular bundle scars per leaf scar. Twigs bright red and lack a rank odor when bruised. This tree occupies habitats ranging from very dry to very wet. It is particularly susceptible to rot following fire or mechanical damage. It tends to sprout from the base of the trunk. *Aceraceae.*

TREES

Box Elder : Leaf scar chevron-shaped. Buds hairy, white. p. 7
Green Ash : Leaf scar straight across top. Seed ½ fruit-length. p. 8
Pumpkin Ash : Leaf scar straight across top. Twigs dull, gray, hairy. Pith edged by orange wood. p. 8
White Ash : Leaf scar notched. Twig smooth. Seed 1/3 fruit-length. p. 8
Blue Ash : Leaf scar notched. Twig four-sided. p. 9
Princess Tree : Leaf scar elliptical. Rattle-box fruits. Gold, tomentose flower buds. ***(See Whorled Buds)*** p. 2

SHRUBS

Fringe Tree : Twigs dull, gray, warty. Leaf scar sunken. p. 9
Coral Berry : Fruits coral-colored. p. 9
Wahoo : Twigs purple, angled. Strong odor. p. 10
Southern Arrowwood : Twigs dark, angled. Pith caramel-brown. p. 10
Rusty Blackhaw : Twigs scurfy. Buds dark brown. p. 10
Shonny Haw : Terminal bud long-pointed. Fruits blue, terminal. p. 11
Buttonbush : Die-back. Leaf scars elliptical. Stipules black. ***(See Whorled Buds)*** p. 2

SMALL TREES

Flowering Dogwood : Twigs purple. Flower buds globose. p. 11
Roughleaf Dogwood : Twigs dull red. Stems light-gray. Pith brown. Uplands. p. 11
Stiff Dogwood : Twigs dull red. Stems gray. Lowlands. p. 12
Silky Dogwood : Twigs dark red with cinnamon hairs on twigs and buds. Pith tawny. Uplands. p. 12

BOX ELDER — *Acer negundo.* Terminal bud imbricate and hairy. Collateral buds opposite, enclosed by common valvate scales. Buds covered with a dense, white, pubescence. Narrow chevron-shaped leaf scars are connected by short, lanceolate, stipule scars. Twigs smooth, lustrous, and green. A tree of moist soils and associated with sugarberry, elm, and ash communities. *Aceraceae.*

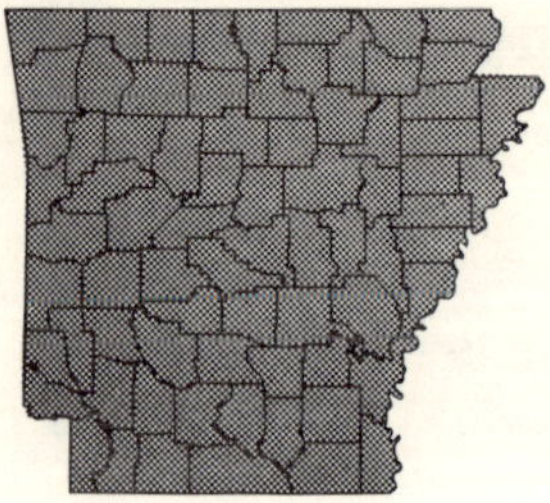

GREEN ASH — *Fraxinus pennsylvanica.* Terminal bud obtuse with three pairs of rufous, tomentose scales. Leaf scars opposite, broad, and straight across the top. Twigs dull and velvety. Winged fruits occurring in clusters on new wood. Seed portion of the fruit extremely narrow and pointed, one-half the length of the fruits. *Oleaceae.*

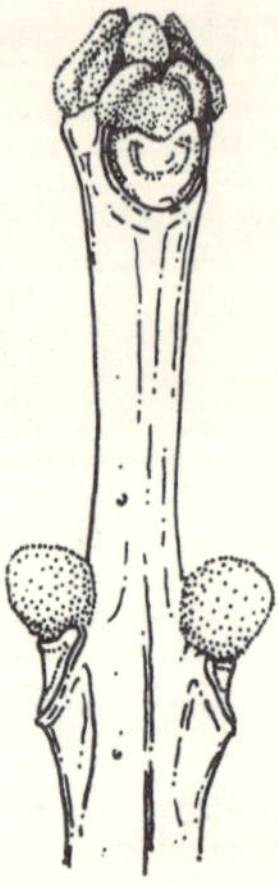

PUMPKIN ASH — *Fraxinus profunda.* Twigs dull gray by short matted hairs. The same kind of tomentum dulls the rusty brown bud scales. Buds valvate with outer and inner fleshy scales, blunted at the tip. Buds short and plump. Lateral buds often superposed and ridged. Leaf scar broad, half-elliptical, broader at the top than bottom because of a flair that hooks out and down. A center longisection shows the pith bordered by a narrow band of orange-gold wood. A slender tree of swamps and wet woods. *Oleaceae.*

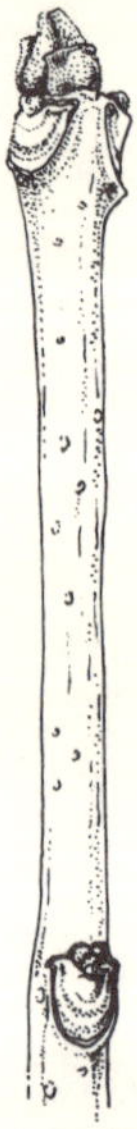

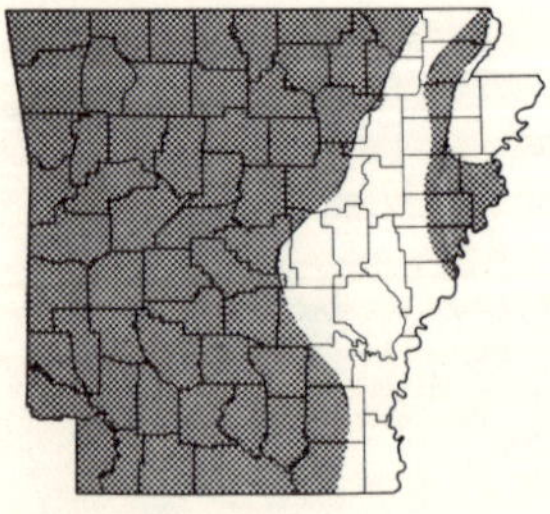

WHITE ASH — *Fraxinus americana.* Terminal bud acute and with four pairs of scales. Leaf scars opposite, broad, and deeply notched across the upper edge. Twigs large, blunt, smooth and glabrous. Winged fruits occurring in clusters on new wood. Seed portion of fruit broadly tapered and one-third the length of the fruit. Large trees of floodplains and moist sites. *Oleaceae.*

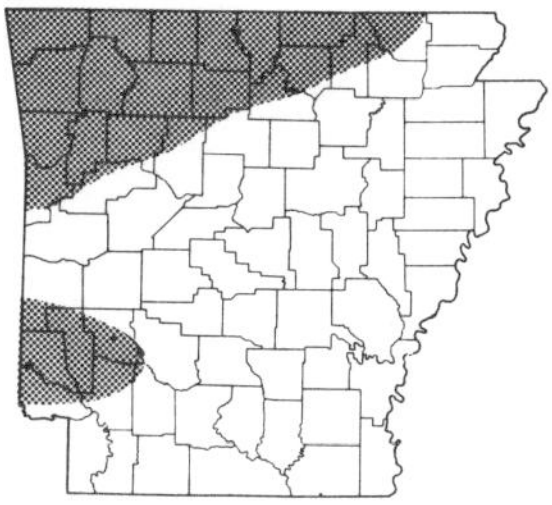

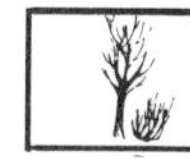

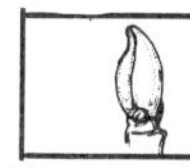

BLUE ASH — *Fraxinus quadrangulata.* Twigs four-sided with corky ridges that arise from each side of the opposite leaf scars and continue down the stem to the next node. These ridges persist on older stems as dark lines on the dull gray smooth bark. Stems pithy. Terminal and lateral buds are valvate and fawn-colored. Leaf scars are round and notched at the top. Vascular bundle scars form a curved line. Trees of limey soils. *Oleaceae.*

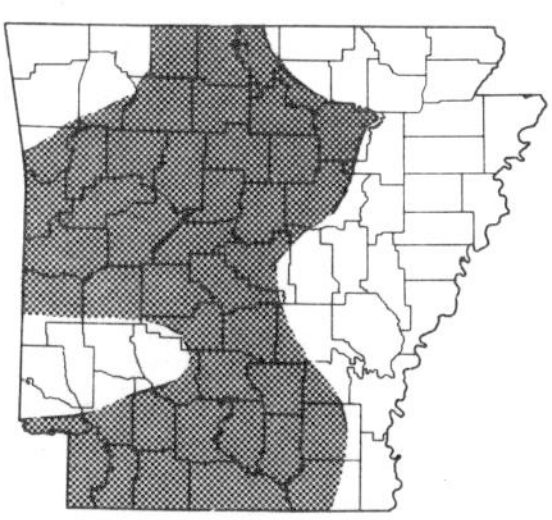

FRINGE TREE — *Chionanthus virginicus.* Terminal bud valvate. Lateral buds opposite. Leaf scar half-elliptical with a prominently rimmed edge. The sunken leaf scar shows a circle or central patch of vascular bundle scars. Twigs dull gray with warty lenticels. A large diffuse shrub of well-drained soils that ranges from rimrock areas, to rocky streams, to upland woods. "Old-Man's-Beard" and "Grancey-Gray-Beard" are other common names used by those transplanting it into yards for ornamental purposes. *Oleaceae.*

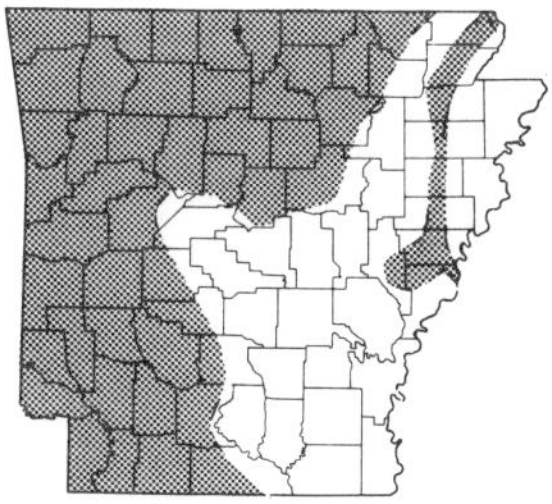

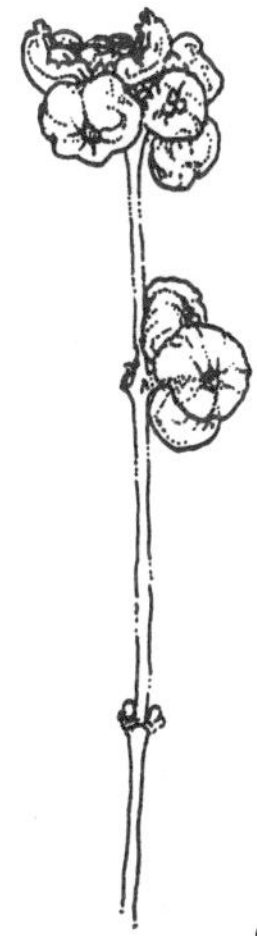

CORAL BERRY — *Symphoricarpus orbiculatus.* The "buckbrush" of woodland pasture and open, dry woods. Thin, limber stems show opposite buds and clusters of over-wintering, coral fruits.
Caprifoliaceae.

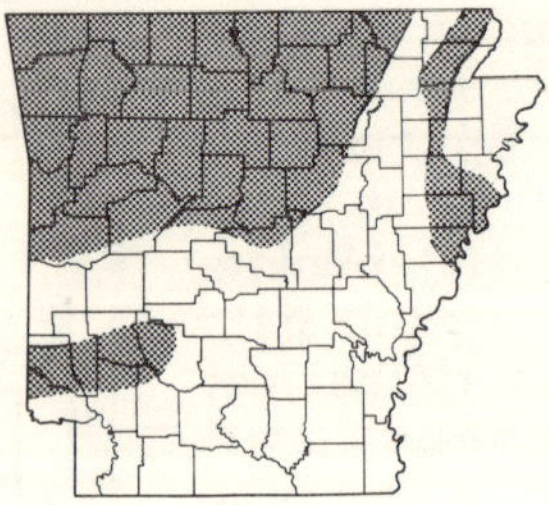

WAHOO — *Euonymus atropurpureus.* "Deep purple" is the meaning of the species name and is descriptive of the winter twig color. Twigs are flat or angular. Older stems are four-angled and twisted. Terminal buds and lateral buds are valvate by two sets of scales. Lateral buds opposite, often sub-opposite, above half-round leaf scars. A single, raised tuft of vascular bundle scars occurs top center of the leaf scar. A strong odor emits from bruised bark. An erect shrub reaching small tree size in moist woods, ravines, and along streams. Commonly associated with limey soils. *Celastraceae.*

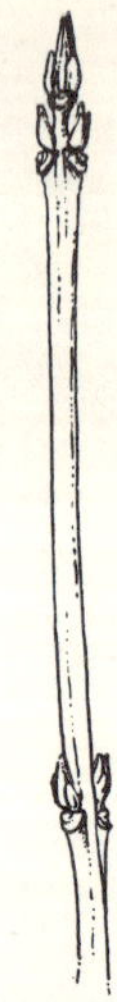

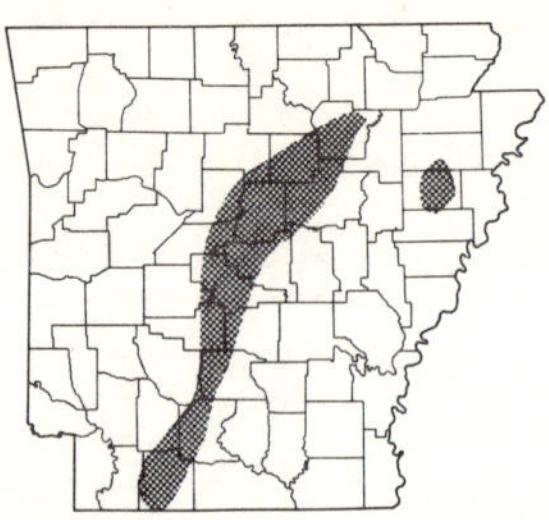

SOUTHERN ARROW-WOOD — *Viburnum dentatum.* Opposite lateral buds thickest near the middle in profile, valvate with two or three sets of scales, with a sharp tip pressed against the twig. The bud appears stalked because of its flattened base. Leaf scar is narrow, chevron-shaped, with each of its three lobes showing a single vascular bundle scar. Twigs smooth and four-angled, ending with an imbricate terminal bud. A large, dark, shrub with long, straight lateral branches arising infrequently along the main stem and then running parallel to it. Pith is a light caramel-brown. Shrubs of stable sandbeds near flowing water. *Caprifoliaceae.*

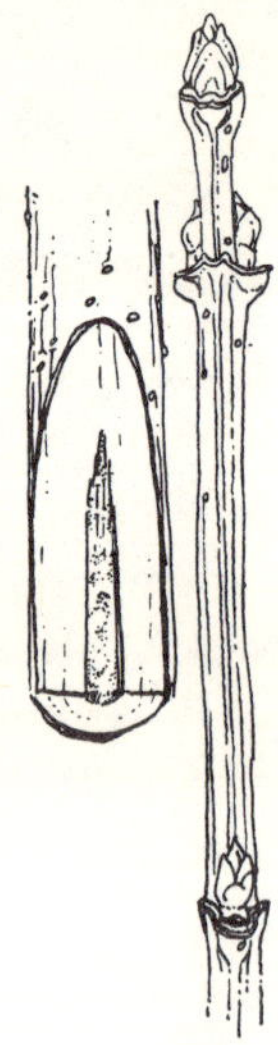

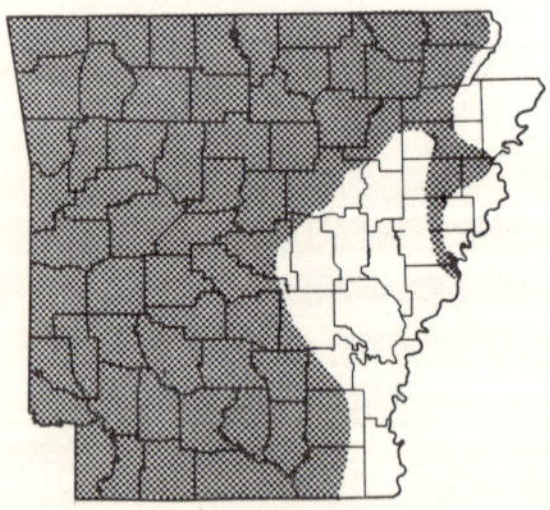

RUSTY BLACK HAW — *Viburnum rufidulum.* Terminal bud present. Lateral buds opposite and valvate. Buds dark brown and tomentose. Twigs scurfy. A small tree, with a diffuse crown found in dry woods and intermittent streams. *Caprifoliaceae.*

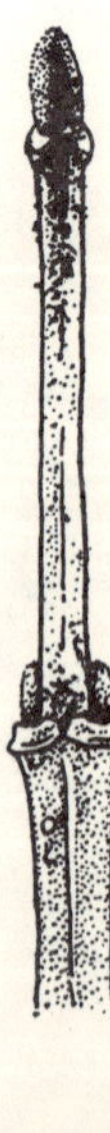

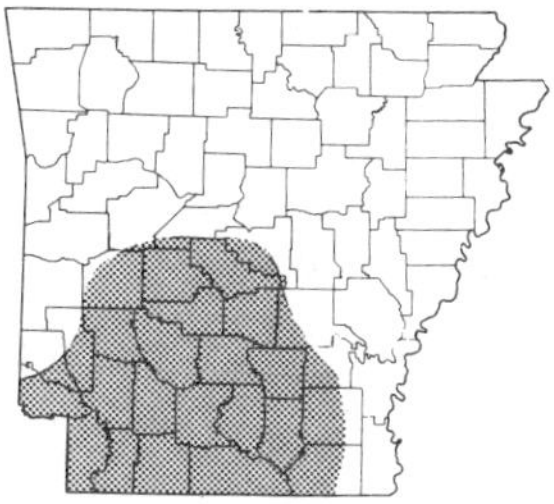

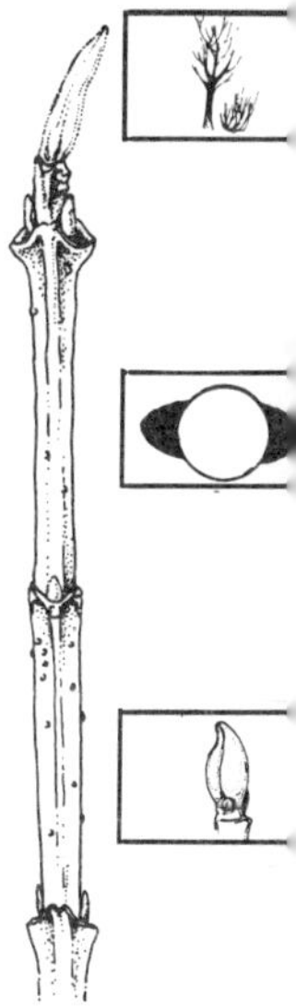

SHONNY HAW — *Viburnum nudum.* Buds and branches opposite. Terminal bud long-pointed, often turned to the side, valvate with two felty or scurfy, red-brown scales. Lateral buds opposite, minute, valvate and flattened against the twig. Leaf scars chevron-shaped, projecting, nearly meeting around the stem, dark with three white groups of vascular bundle scars. Terminal, tight clusters of small numerous blue fruits signal the occurrence of this graceful shrub in the winter woods. Found in sandy springheads, swamps, and areas of deep shade. *Caprifoliaceae.*

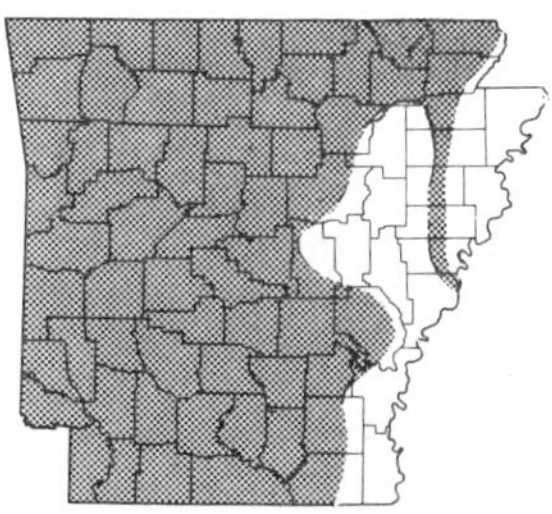

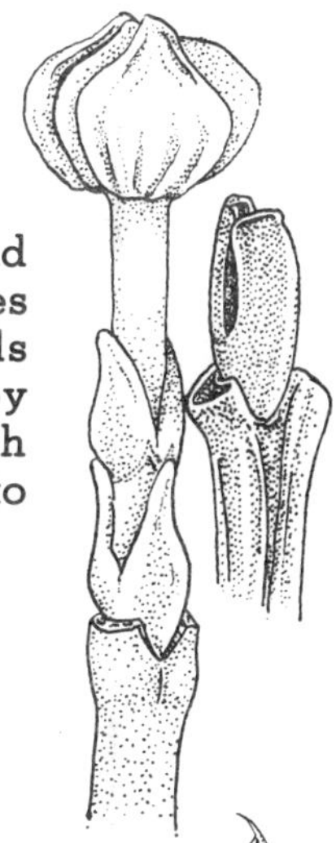

FLOWERING DOGWOOD — *Cornus florida.* Terminal bud valvate and purplish. Lateral buds opposite, minute, and embedded. Twigs purplish and angled. Tips of lateral branches curved and erect. Flower buds terminal, stalked, subtended by bracts, and sub-globose with acute tips. A small under-story tree with bark broken into thin, square plates. *Cornaceae.*

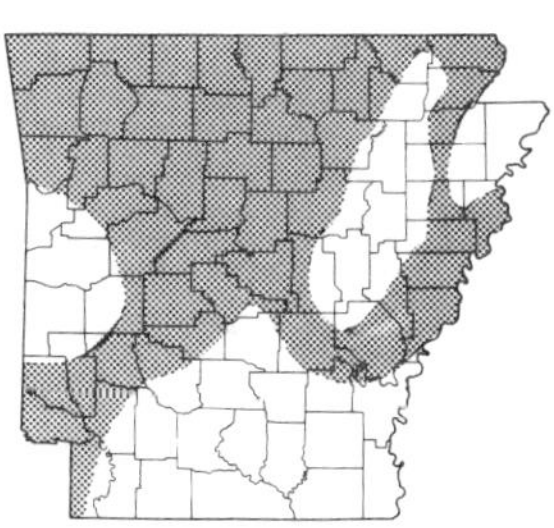

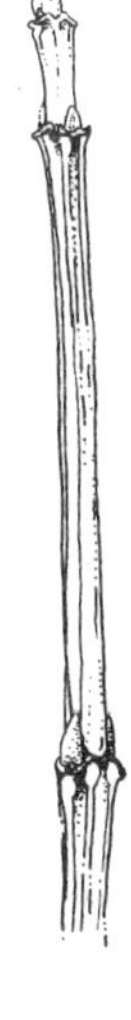

ROUGHLEAF DOGWOOD — *Cornus drummondii.* Tall shrubs or small, slender trees growing in colonies. Dull, hirsute, red twigs and light-gray stems with brown pith are distinct field marks. Valvate terminal buds are dark and hirsute. Opposite lateral buds are valvate, brown and hirsute. Leaf scars are lobed chevrons with a vascular bundle scar at each lobe. Stipular scars connect the leaf scars around the stem. Shrubs of moist, sweet, well-drained slopes, ledges and ravines, of open upland forests. *Cornaceae.*

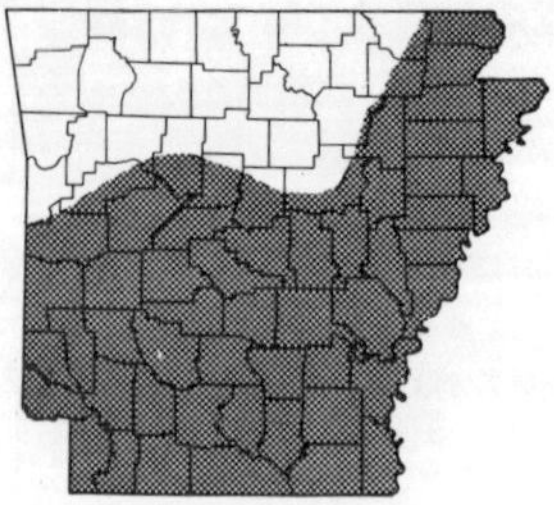

STIFF DOGWOOD — *Cornus foemina* Tall, spreading, gray shrubs with dull red twigs. Found on stream banks in low wet woods and swamps. Terminal bud usually absent with opposite lateral buds projecting forward forming a fork. Forked branching follows as a consistent growth pattern. Lateral buds slender, appressed, pointed and silky with long, flattened, brown hairs. Chevron-shaped leaf scars have three vascular bundle scars. *Cornaceae.*

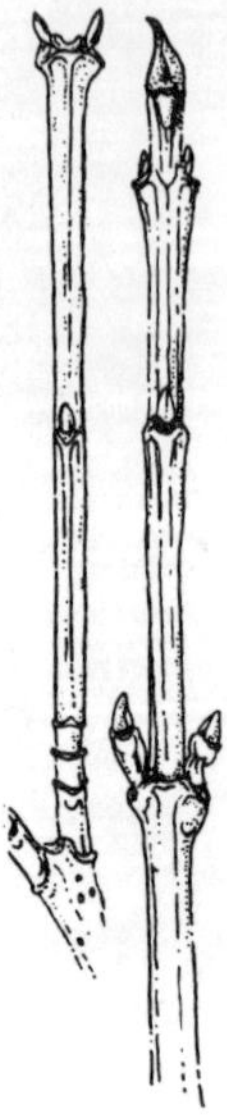

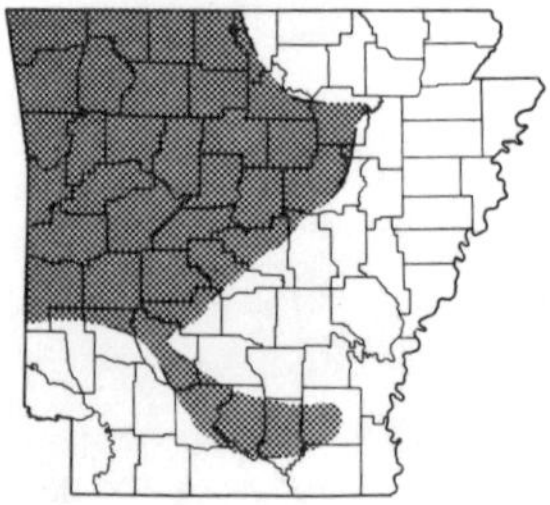

SILKY DOGWOOD — *Cornus amomum* Dark red twigs pubescent with dense, flattened white hairs, becoming cinnamon-colored at the tips. Terminal and lateral buds covered with cinnamon tomentum. Pith tawny. Buds opposite above elevated, crescent, leaf scars showing three, white, vascular bundle scars. An upright and spreading shrub. The "red willow" of upland hardwood ravines. *Cornaceae.*

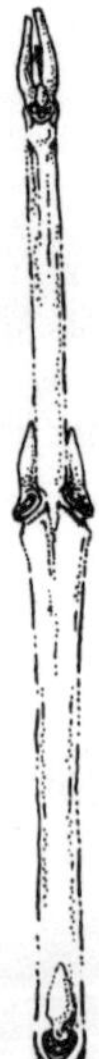

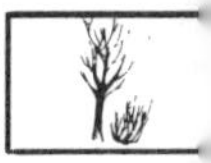

BUDS OF EQUAL SIZE

Beech : A single bud at the twig tip. p. 13

Oaks : Bunched buds at the twig tip. pp. 14-20

- **White Oak Group** : Buds long as broad. Acorn cup smooth inside. Acorn sweet. pp. 14, 15
- **Chestnut Oak Group** : Buds longer than broad. Acorn cup smooth inside. Acorn sweet. p. 16
- **Red Oak Group** : Buds longer than broad. Acorn cup wooly inside. Acorn bitter. pp. 16-19
- **Willow Oak Group** : Buds longer than broad. Number of bunched buds reduced. Acorn cup wooly inside. Acorn small, bitter. pp. 19,20

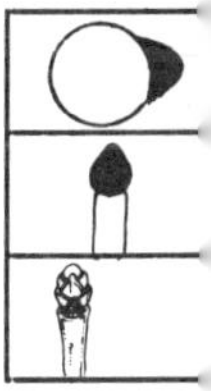

BUDS OF UNEQUAL SIZE

Poplars : Stipular scars. Obtrangular leaf scars. pp. 20, 21

Fruit Trees and Relative : Buds usually hairy, loosely imbricate. Buds strongly colored. Fruit spurs common. pp. 21-23

Corkwood : Streaked tomentum. Light-weight wood. p. 24

Chinese Parasol Tree : Leaf scar broad. Terminal bud spherical. Felty. p. 24

Shagbark Hickory : Leaf scar broad. Terminal bud pointed. p. 24

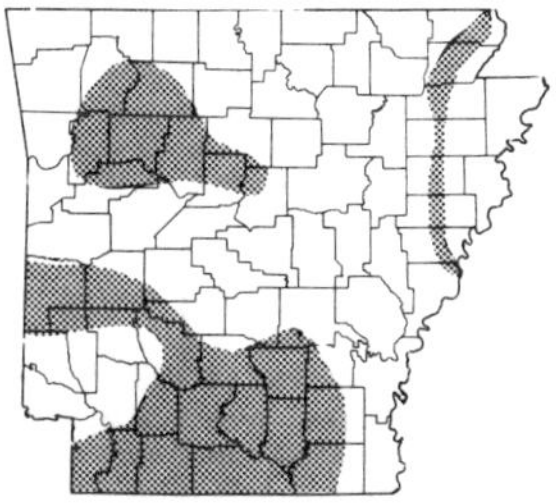

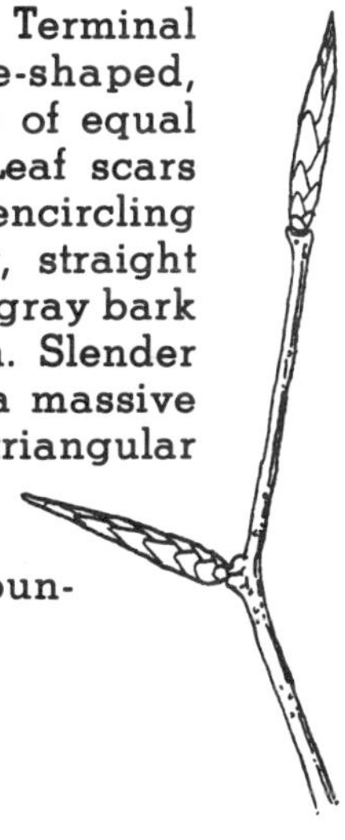

BEECH — *Fagus grandifolia.* Terminal bud imbricate, long, lance-shaped, smooth, and brown. All buds of equal size. Lateral buds alternate. Leaf scars small with stipule scar nearly encircling the thin, stiff twig. The clear, straight trunk has a thin, smooth, blue-gray bark with a faint horizontal pattern. Slender branches spread widely into a massive crown. One can find small, prickly husks containing triangular beechnuts. Found in rich woods and uplands of Crowley's Ridge, along floodplains of the Coastal Plains, and in restricted coves of the Ouachitas and Ozark Mountains. *Fagaceae.*

Summary
OAK BUDS AND ACORNS

Group	Species	Buds Longer Than Broad	Acorn Cup Wooly Inside	Acorn Cup Scales Hairy	Acorn Cup Stalked	Acorn Cup Fringed	Acorn Cup Saucer-Shaped	Acorn Striped-Striated	Acorn Sweet
	White Oak	-	-	+	-	-	-	-	+
White	Bur Oak	-	-	+	+	+	-	-	+
Oaks	Post Oak	-	-	+	-	-	-	+	+
	Overcup Oak	-	-	+	-	-	-	-	+
Chestnut	Cow Oak	+	-	+	-	+	-	-	+
Oaks	Chinkapin Oak	+	-	+	-	-	-	-	+
	Northern Red Oak	+	+	+	-	-	+	-	- yllw
	Black Oak	+	+	+	-	+	-	+	- yllw
	Southern Red Oak	+	+	+	-	-	+	-	- yllw
	Shumard's Oak	+	+	-	-	-	+	-	- wht
Red	Maple-Lvd. Oak	+	+	-	-	-	+	-	- wht
Oaks	Blackjack Oak	+	+	+	-	-	-	+	- yllw
	Scarlet Oak	+	+	-	-	-	-	apical rings	-
	Arkansas Oak	+	+	+	-	+	+	-	-
	Pin Oak	+	+	-	-	-	+	+	- yllw
	Nuttall's Oak	+	+	-	-	-	-	-	- yllw
Willow	Willow Oak	+	+	+	-	-	+	-	-
Oaks	Water Oak	+	+	+	-	-	+	-	-
	Shingle Oak	+	+	+	+	-	-	+	-

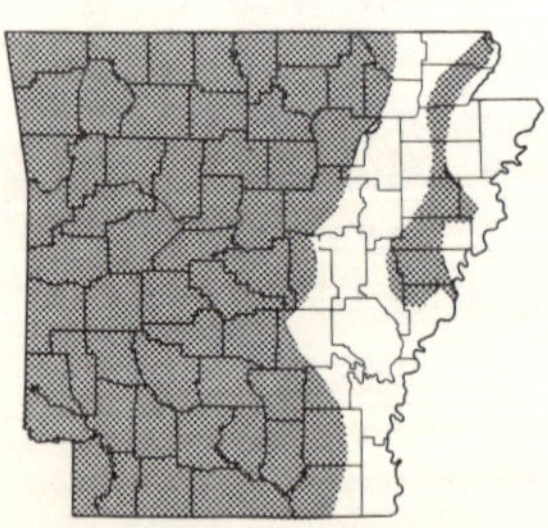

WHITE OAK — *Quercus alba.* (White Oak Group). Terminal bud and several lateral buds clustered at the twig tip. Buds equal sized and with numerous imbricate scales. Buds alternate, dark red, glabrous, and globose. The diameter of the terminal bud base is equal to or three-fourths of the bud length. Leaf scars obtriangular with rounded tips. Twigs lustrous, silvery white to purple with prominent lenticels. Branches with light gray, scaley, bark. A tree growing in nearly all moist upland sites restricted only by arid ridges and poorly drained bottomland. *Fagaceae.*

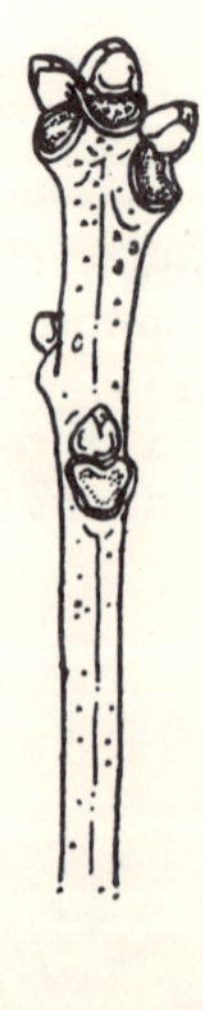

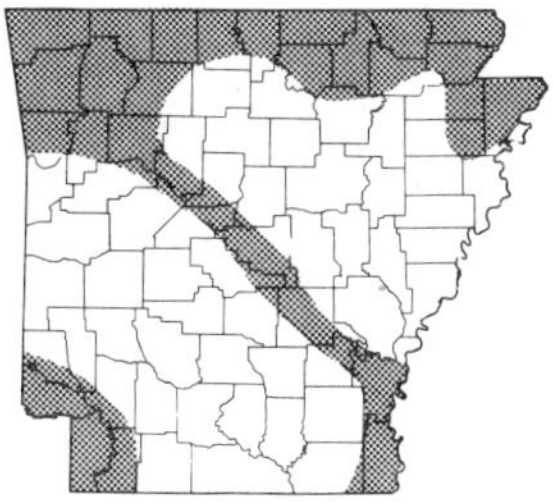

BUR OAK — *Quercus macrocarpa.* (White Oak Group). Terminal bud and several lateral buds bunched at the tip of a coarse, blunt twig. Buds imbricate, alternate, dark mahogany with pale tomentum, nearly broad as long. Leaf scars are lobed crescents often attended by withered stipules. Twigs pale with soft, finely-furrowed bark. Warty lenticels of twigs develop into corky ridges in older stems. Bark of trunk thick and deeply furrowed. One can find large "mossy cup" acorn caps under old trees. Restricted to habitats of high floodplains and well-drained bottomland. *Fagaceae.*

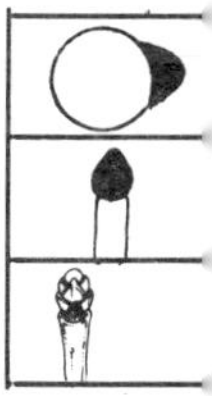

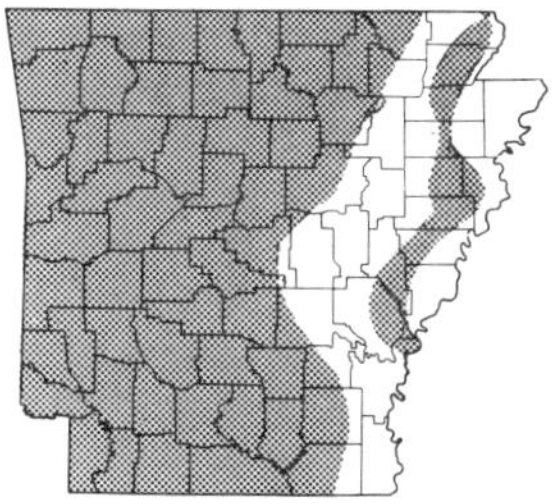

POST OAK — *Quercus stellata.* (White Oak Group). Terminal bud and a few lateral buds clustered at the twig tip. Buds alternate, globose, and imbricate with red, ciliate scales. Leaf scars elevated, reniform, and dark compared to the yellowish tomentose twig. Branches appear dark and shaggy. Bark of trunk is light gray, finely furrowed and tight. Arid ridges and soils alternately water-logged and dry are its characteristic sites. *Fagaceae.*

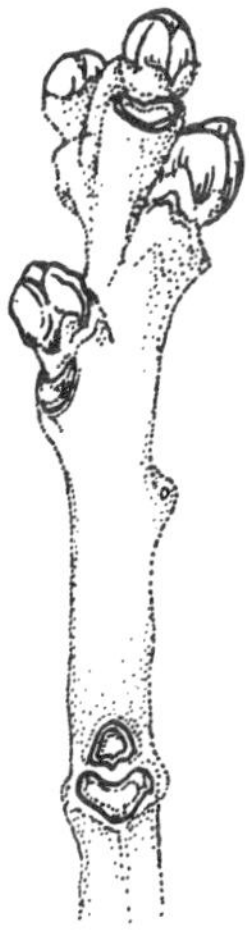

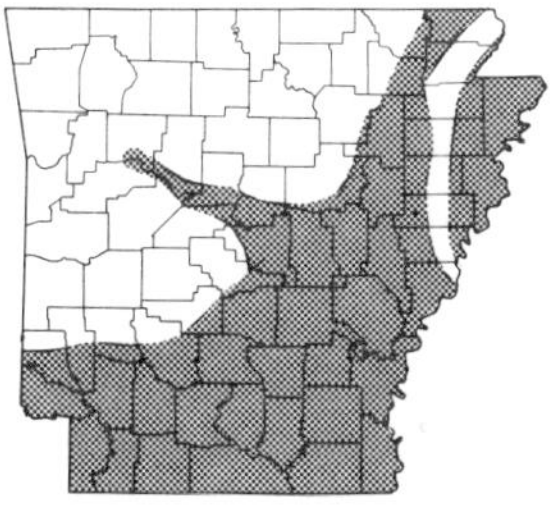

OVERCUP OAK — *Quercus lyrata.* (White Oak Group). Terminal bud and a few lateral buds clustered at the tip of a blunt twig. Lateral buds alternate, imbricate, deltoid, and brown. Leaf scars elevated, reniform, dark with white vascular bundle scars. Stiff, erect, stipules persist into winter. Twigs gray-green or red with a silvery sheen. Find acorns with fringed, deep caps. Overcup Oak is tolerant of long periods of flooding in backwater areas and poorly drained sites of bottomland. *Fagaceae.*

COW OAK — *Quercus michauxii.* (Chestnut Oak Group). Terminal bud attended by two or three lateral buds. Buds imbricate, reddish-brown, and four-sided. Bud scales thin, sparsely hairy with pale margins. Bark thick, ash-gray, with irregular furrows and narrow ridges. Often with white oak but extending beyond its range to wetter habitats. *Fagaceae.*

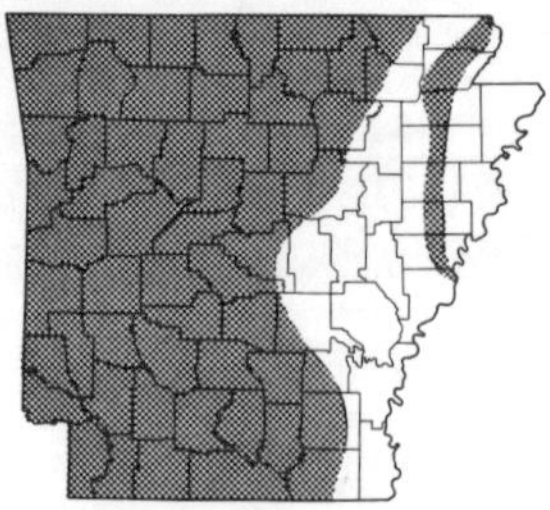

CHINKAPIN OAK — *Quercus prinoides var. acuminata.* (Chestnut Oak Group). The prominent terminal bud is accompanied by two or three lateral buds. Buds imbricated and dark orange. Twigs slender, ridged, angled, glabrous and dark reddish-brown to gray. Bark ash gray often with red tinges, weathering into thin scales. Found in mixed hardwood forests on slopes where soils are deep, rich, and moist. *Fagaceae.*

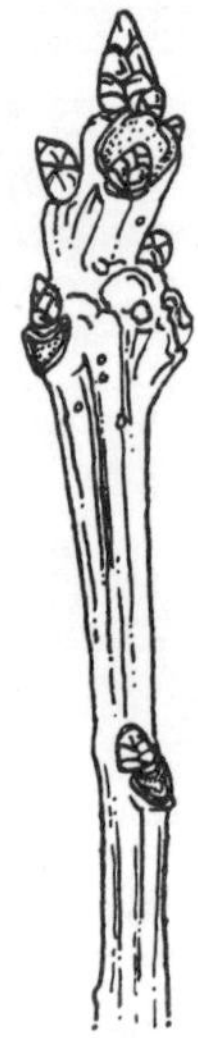

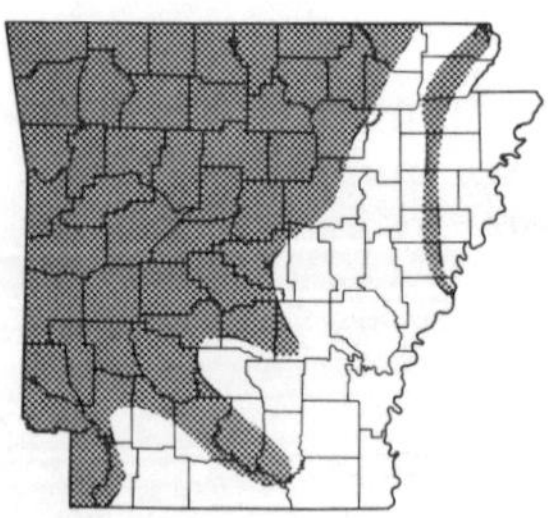

NORTHERN RED OAK — *Quercus rubra.* Bunched buds at the twig tip, imbricate and reddish-brown. Bud scales hairy or downy at the tip; ciliated. Twigs stout and shiny. Limbs of the crown and the high trunk show bark with shiny, gray plates alternating with thin, black fissures. This streaking marks the "zebra of the winter woods". A knife-cut to the inner bark shows light red tissue. A tree of north slopes. *Fagaceae.*

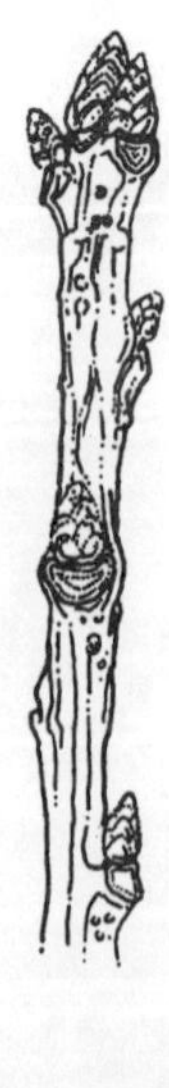

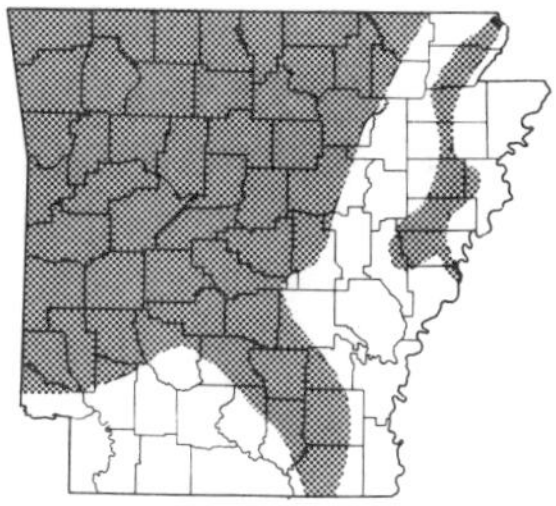

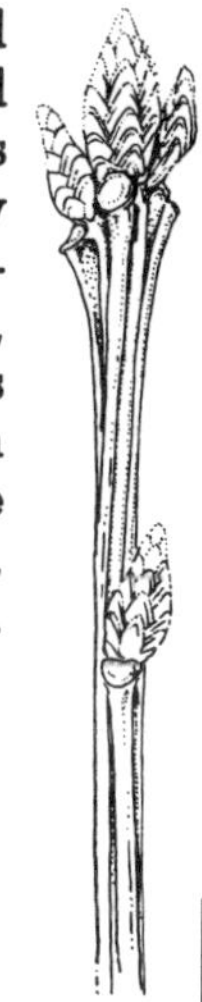

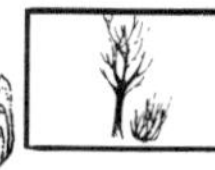

BLACK OAK — *Quercus velutina.* (Red Oak Group). Terminal bud with lateral buds bunched at the twig tip. Buds sharp-pointed, angled, with gray downy, imbricate scales. Leaf scars half-round and lopsided. The twigs are stout, angular, and glabrous. The bark is black, lacks lustre, deeply furrowed with dull, scaly ridges. A cut made to the inner bark reveals a bright orange-yellow pulp, bitter to the taste. A mountain tree of drier sites and mixed hardwood forests. *Fagaceae.*

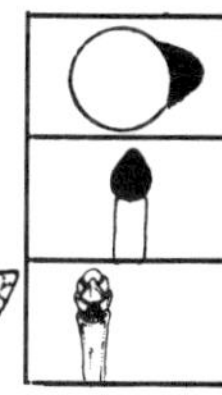

SOUTHERN RED OAK — *Quercus falcata.* (Red Oak Group). Terminal bud and a few lateral buds clustered at the twig tip. Buds alternate, longer than broad, plump, round in cross section, imbricate, shiny, and bright reddish-brown. leaf scars half-round, dark and with scattered vascular bundle scars. Twigs coarse, pubescent, and reddish-brown. Branches pale gray and mottled. Bark of old trees black, massive with broad ridges and deep fissures. Inner bark with a pale salmon color. Grows in dry upland sites. *Fagaceae.*

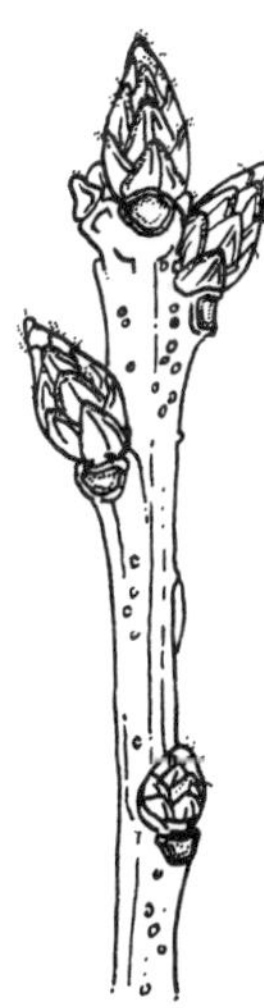

SHUMARD'S OAK — *Quercus shumardii.* (Red Oak Group). Terminal bud and several lateral buds clustered at the twig tip. Buds imbricated with ciliate or downy scales, straw-colored, four-angled, alternate, and longer than broad. Leaf scars half-round with scattered vascular bundle scars. Twigs glabrous, bright olive-green with elongate, white lenticels. Found high on slopes of moist ridges. *Fagaceae.*

MAPLE-LEAVED OAK — *Quercus shumardii* var. *acerifolia.* Differing from Shumard's Oak by the dull ash-gray bark of the twig and the smooth margins of the bud scales. Acorns sought after and gleaned by rodents early in the mast season. A small multi-trunked tree of cliffs and high coves on Mt. Magazine. Rare. *Fagaceae.*

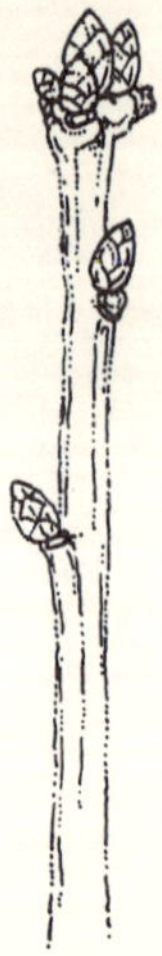

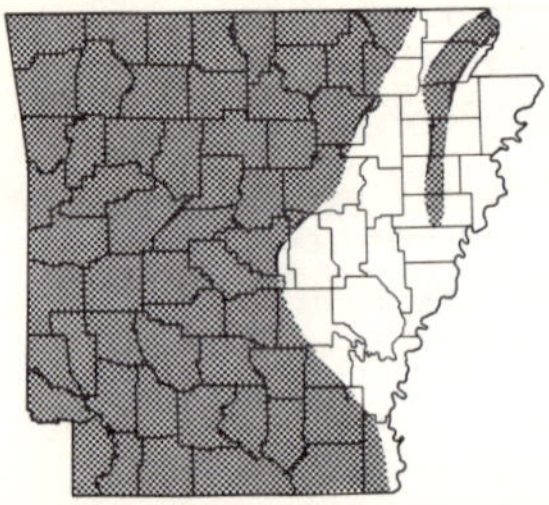

BLACKJACK OAK — *Quercus marilandica.* (Red Oak Group). The wooly, tawny, terminal bud is accompanied by one or two lateral buds at the twig tip. Bud alternate, imbricate, long-pointed, with a felty tomentum. Leaf scars prominently elevated and widely obovate. Twigs coarse, irregular, dull, and scurfy. Limbs contorted and branched irregularly. Bark black, hard, and furrowed into blocks. A common tree of sterile, arid, soils. Dominates in savannas caused by repeated fires. *Fagaceae.*

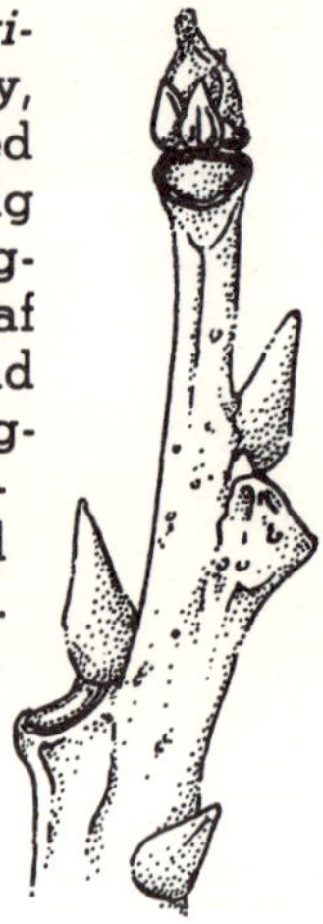

ARKANSAS OAK — *Quercus arkansana.* Twigs slightly pubescent with short internodes, numerous lateral buds and two or three buds clustered at the tip. Second-year stems darker and smoother. Buds short, rounded at the base and coming to an abrupt point. Bud scales thin, light brown, with membraneous margins. The bark is black, thick, narrowly ridged and with tight scales. Growth form shows a narrow crown with stout, drooping limbs. *Fagaceae.*

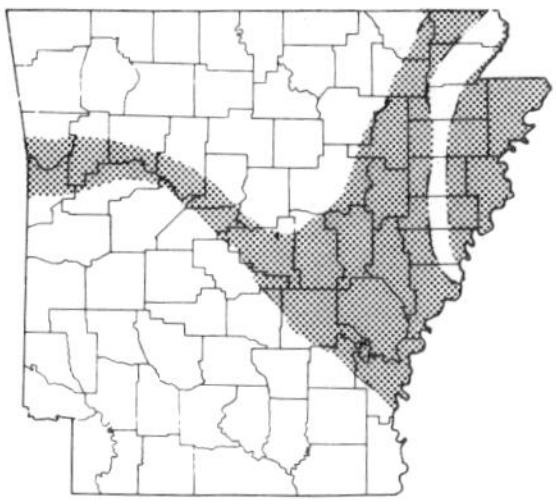

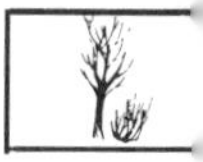

PIN OAK — *(Quercus palustris).* (Red Oak Group). Buds clustered around the terminal bud. Buds imbricate with ciliated bud scales, reddish-brown and flattened. Leaf scars alternate, elevated, and widely obovate. Twigs slender, lustrous, and reddish-brown. Strong terminal branch growth with many slender side branches. Leaves often persist into winter. In floodplains, backwater, and swamps with tupelo gum, cypress, and overcup oak. *Fagaceae.*

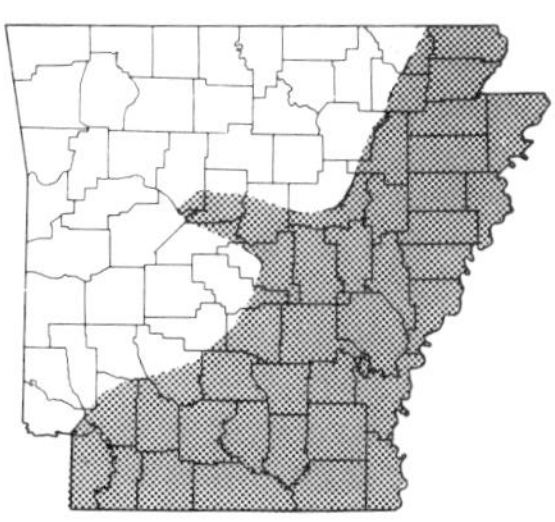

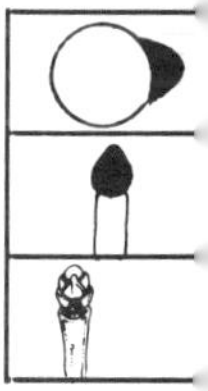

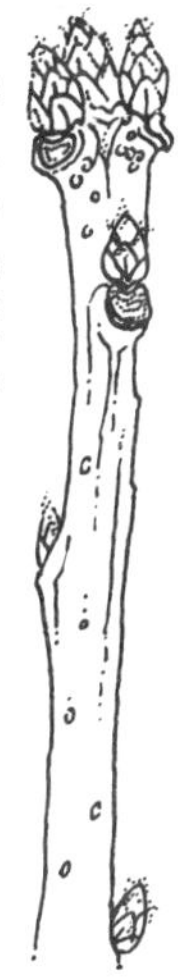

NUTTALL'S OAK — *Quercus nuttallii.* (Red Oak Group). Lateral buds alternate and clustered at the tip with the terminal bud. Buds imbricate and chestnut-brown. Bud scales with white ciliate margins. Leaf scars elevated and half elliptical. Twigs red to brown with white lenticels. Bark of old trees thin, dark, slightly fissured, grading out to a smooth slate gray up the trunk. Grows on overflow banks of sloughs and rivers. *Fagaceae.*

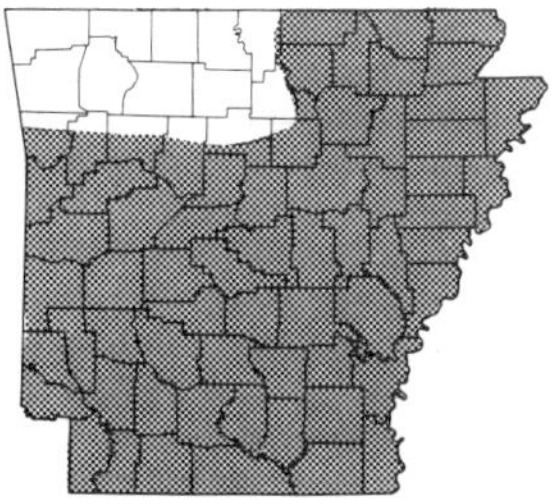

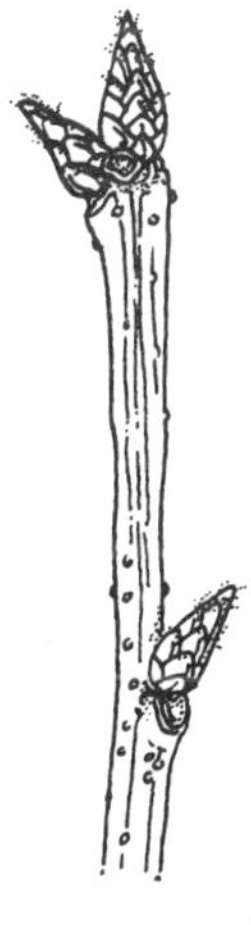

WILLOW OAK — *Quercus phellos.* (Willow Oak Group). Terminal bud associated with one or two lateral buds at the twig tip. Buds imbricate, sharp, longer than broad, distinctly four-angled. Bud scales glabrous, reddish-brown and pale at the borders. Leaf scars alternate, and half elliptical. Twigs gray-green, delicate, and shiny, with diameters up to two millimeters. A straight, massive bole and a diffuse crown of delicate branches characterize its mature growth form. Dominates with even-aged populations in the "pin oak flats" of the Arkansas lowlands. *Fagaceae.*

WATER OAK — *Quercus nigra.* (Willow Oak Group). Terminal bud and lateral buds longer than broad, imbricated with numerous scales, four-angled, chestnut brown and downy white at the tips. White-tipped buds conspicuous in reflected sunlight. Leaf scars alternate and half elliptical. Twigs dull, brown to gray, glabrous with warty lenticels. Branches with strong terminal growth and numerous short branches. Bark smooth gray finally breaking into shallow furrows on old trunks. In overflow bottomland or floodplain terraces. *Fagaceae.*

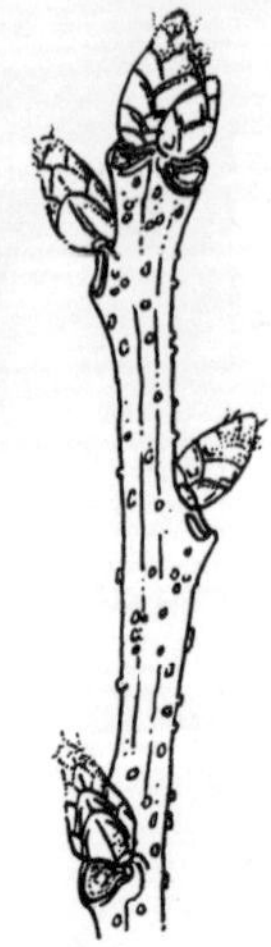

SHINGLE OAK — *Quercus imbricaria.* (Willow Oak Group). Twigs and buds a dull chestnut-brown often with persistent leaves or leaf petiole stubs. Terminal bud and three to four lateral buds clustered at the twig tip. Buds longer than broad, imbricated with thin, ciliate scales. Older branches with a gray sheen on the smooth chestnut-brown bark. On high terraces of streams and slopes in north Arkansas. *Fagaceae.*

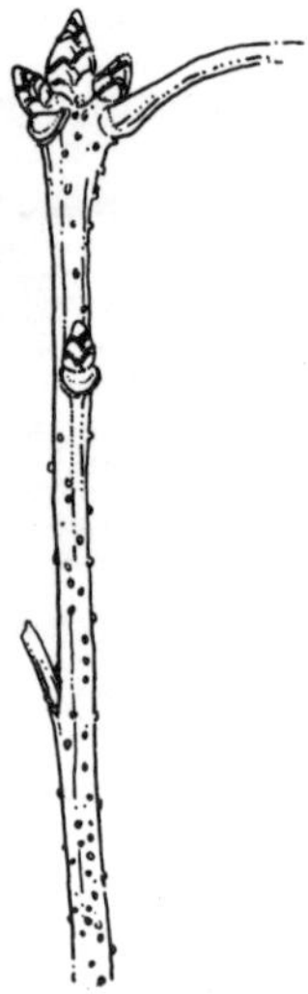

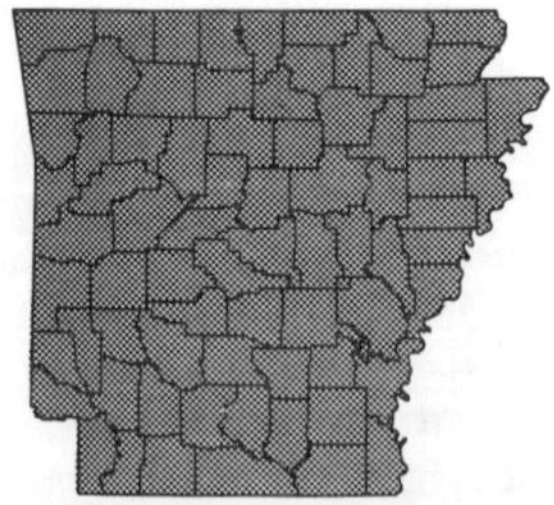

COTTONWOOD — *Populus deltoides.* Terminal bud imbricate, sharp-pointed, and resinous when crushed. Lateral buds alternate. Buds and twigs glabrous. Stipular scars short. Leaf scars obtriangular with three vascular bundle scars. Twigs ridged and irregular in cross section. Lateral branches commonly self-pruning and accumulating on the ground. Twigs dull yellow. Branches smooth and yellow-green. Bark of old trunks thick with flat, smooth ridges and narrow, deep fissures. Large, vigorous, fast-growing trees sprouting freely from stumps and roots. Growing along sluggish streams, water courses, and in the flood plains of major streams. *Salicaceae.*

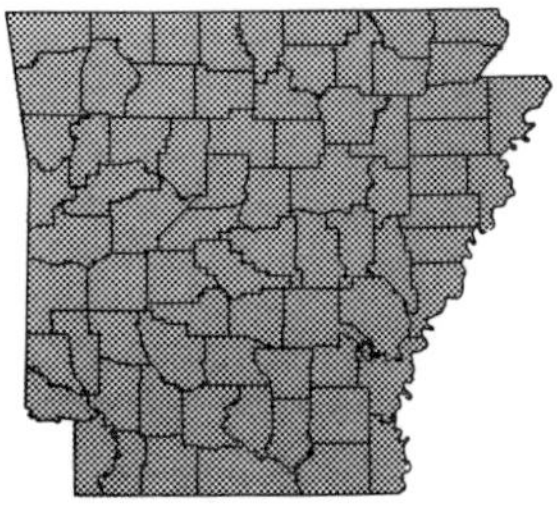

LOMBARDY POPLAR — *Populus nigra* var. *italica.* A tall extremely slender tree, striking in appearance because of its formal, columnar growth habit. Branches are erect and growing parallel to the trunk. Twigs yellow, branches shiny and ash-gray. Terminal buds imbricate, brown and green, its length thrice its width at the base, aromatic, and resinous when crushed. Lateral buds alternate, long-pointed. Leaf scars broadly obtriangular with three vascular bundle scars. Stipular scars short. Introduced for cultivation and widely propagated by hardwood cuttings. *Salicaceae.*

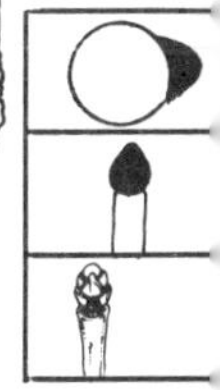

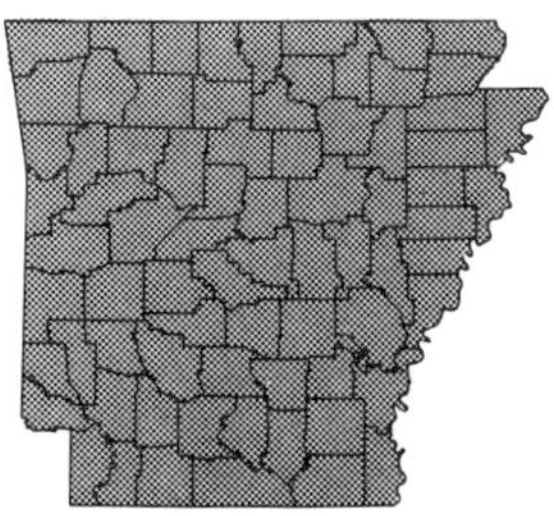

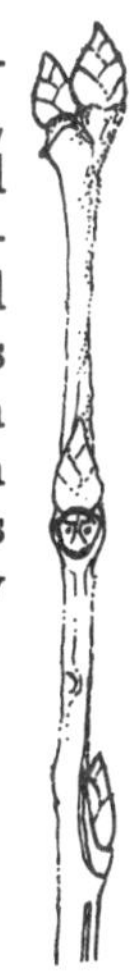

SILVER POPLAR — *Populus alba.* Terminal buds acutely pointed, imbricate, and tomentose with white hairs. Lateral buds alternate on tomentose twigs. Stipule scars short. Older stems become dull gray with conspicuous black lenticels arranged in horizontal patterns. Will form thickets by root sucker clones. It has been introduced for cultivation and spreads easily. Subject to damage and death by severe winters. *Salicaceae.*

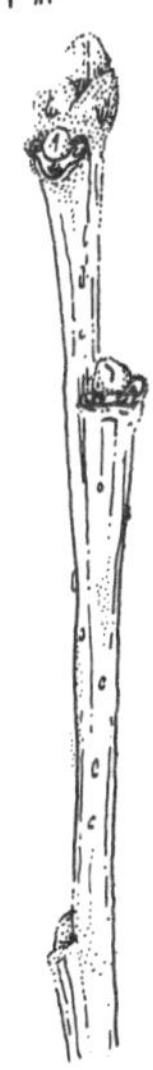

APPLE — *Pyrus malus.* Buds and twigs with a dense, irregular, gray feltiness. Terminal bud imbricated. Lateral buds alternate above a three-lobed, chevron-shaped leaf scar with three vascular bundle scars. Twigs often dark red; branches smooth, shiny, olive-colored; trunks with a smooth gray bark. Water sprouts common on low angled limbs of the spreading crown. Introduced, widely cultivated, and occasionally escaping. *Rosaceae.*

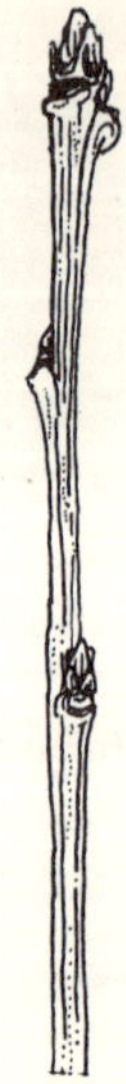

PEACH — *Prunus persica.* Small, broad-crowned trees with bright green striated twigs and red, sharp-pointed, appressed buds. Bruised twigs have a rank, bitter-almond taste and odor. Introduced, cultivated and escaping. Common to old house-sites, abandoned orchards, and old barn lots. Peach "mummies," an indicator of the brown-rot disease, hang among the branches as a usual winter aspect of feral trees. *Rosaceae.*

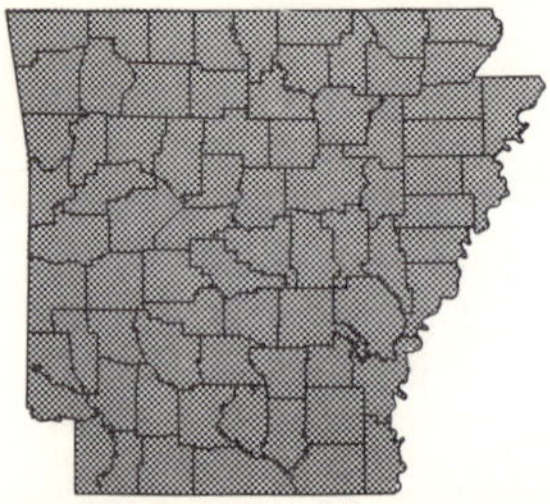

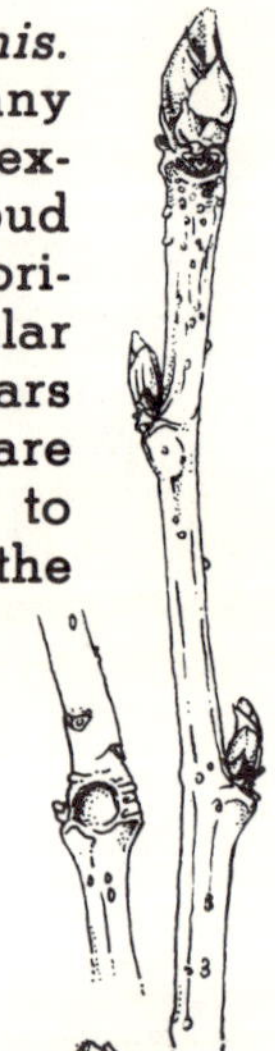

COMMON PEAR — *Pyrus communis.* Terminal bud imbricated with mahogany colored, shiny scales; its diameter exceeds that of the twig and is half the bud length. Alternate lateral buds are imbricate above small, narrow, obtriangular leaf scars. Three vascular bundle scars appear in each leaf scar. Twigs are smooth and shiny, reddish-brown to olive, with scattered lenticels. Broad, flat, fruit scars occur in the crotches of young branches. Branches from lateral buds develop into a profusion of main trunks that finally spread and droop strongly at the edge of the crown. Widely cultivated but also found away from house and orchard sites. *Rosaceae.*

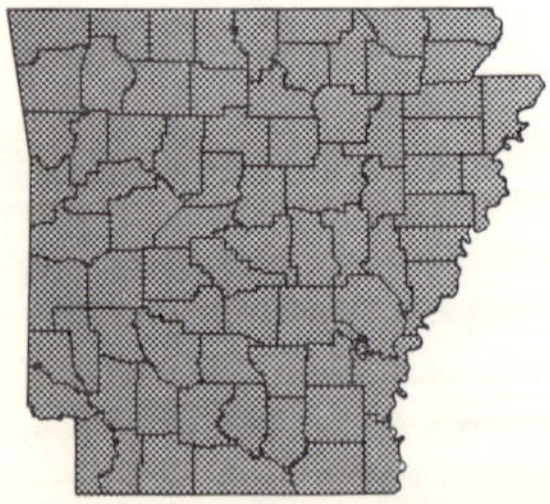

FLOWERING CRAB — *Pyrus floribunda.* Terminal bud loosely imbricated with tomentose scales. Lateral buds alternate above crescent-shaped or half-round leaf scars. A line of vascular bundle scars shows along the central leaf scar. Similar to apple, with numerous fruit spurs on two- and three-year-old branches. A small tree with a thick crown and arching branches that become purple with exposure to bright sunlight. Introduced from Japan. *Rosaceae.*

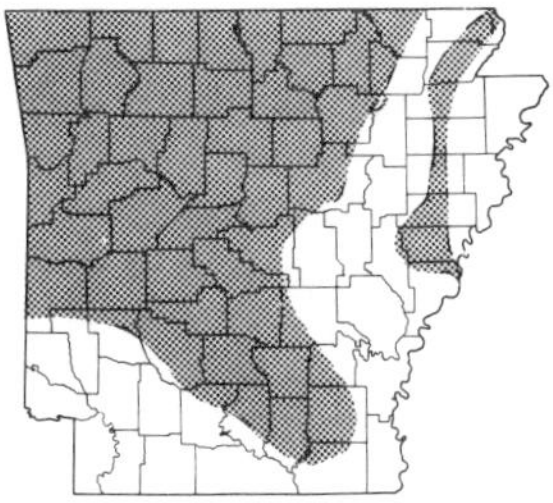

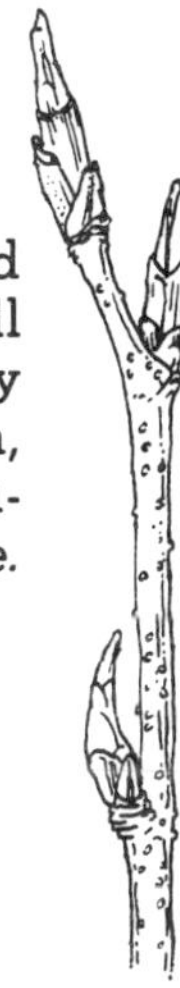

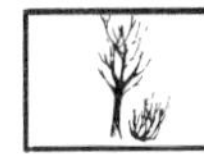

SHADBUSH — *Amelanchier arborea.* Terminal bud imbricate, long, pointed, reddish-brown and often with harlequin color patterns: alternating colors of red, brown, yellow, and green bud scales. Lateral buds alternate with small and inconspicuous leaf scars. A sparsely branched, small, slender tree with a thin, gray bark, scattered among upland timber and on dry mountain slopes. *Rosaceae.*

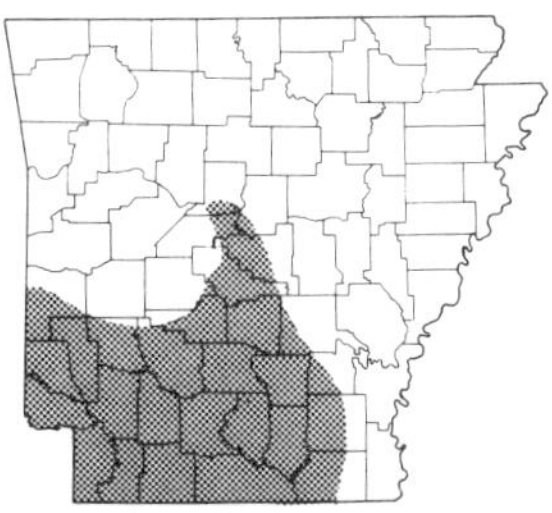

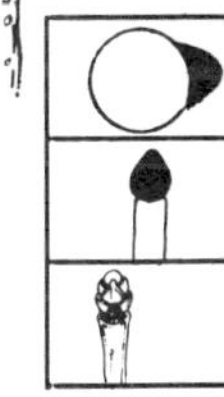

RED CHOKEBERRY — *Pyrus arbutifolia.* Terminal buds pink, long-pointed, and imbricate, contrast with the thick, gray tomentum of the slender twigs. Twigs feel soft from the covering. Alternate lateral buds are imbricate, glabrous, often oblique. Leaf scars are inconspicuous, thin lines showing three patches of vascular bundle scars. Old branches are slick, shiny, gray mottled, astringent and bitter. A clumpy shrub with slender branches, growing in sandy springheads, swamps, and boggy woods. Find sparse, thin clusters of small red fruits persisting into winter. *Rosaceae.*

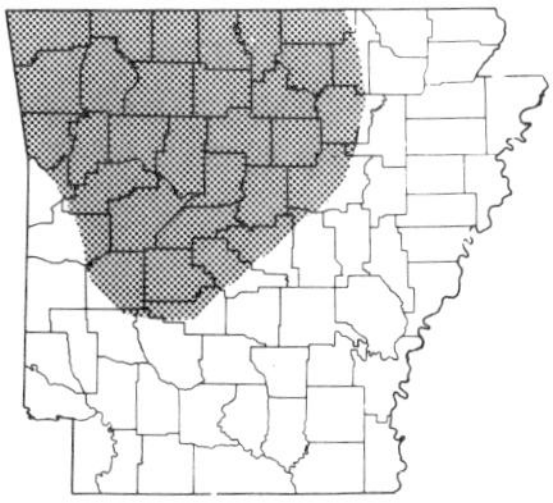

NINEBARK — *Physocarpus opulifolius.* Dense shrubs of bluffs, rocky slopes and stream beds. Twigs smooth, dull, light brown, with downy, imbricated buds and strongly decurrent internodal ridges originating at two corners of the raised obtriangular leaf scar. Older bark shreds away revealing numerous layers; a conspicuous feature of shrubs in thickets. *Rosaceae.*

CORKWOOD — *Leitneria floridana.* The red-brown twigs are tomentose in streaks alternating with smooth glossy bark. Light-colored elongated lenticels erupt through the bark. The terminal bud and flower cluster aments are covered with a gray tomentum. Lateral buds alternate, imbricate, tomentose, and above obtriangular to half-round leaf scars. Find three vascular bundle scars. A wetlands shrub that spreads from clumps by cloning. The wood is lighter than cork. *Leitneriaceae.*

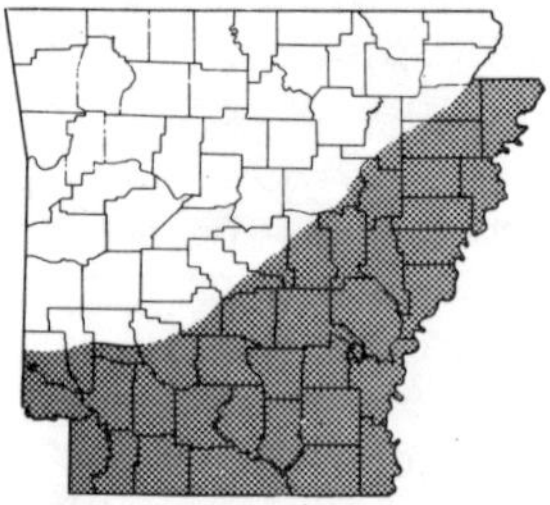

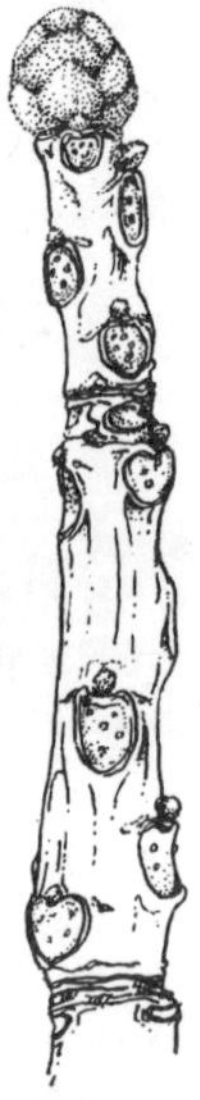

CHINESE PARASOL TREE — *Firmiana platanifolia.* A large, spherical, imbricate, brown, felty bud terminates stout dark-green twigs. The lateral buds are similar but a fraction of the size of the terminal bud. Leaf scars circular to elliptical, rimmed and saucer-shaped, with four to five vascular bundle scars randomly scattered in the center. Faint stipular scars are present. Sparse, long, forked stems develop a tall, thin-crowned tree. The bark is smooth and gray. Remnants of flower clusters resemble the ribs of a stripped, upturned parasol. Introduced, cultivated and escaping in southern Coastal Plains regions. *Sterculiaceae.*

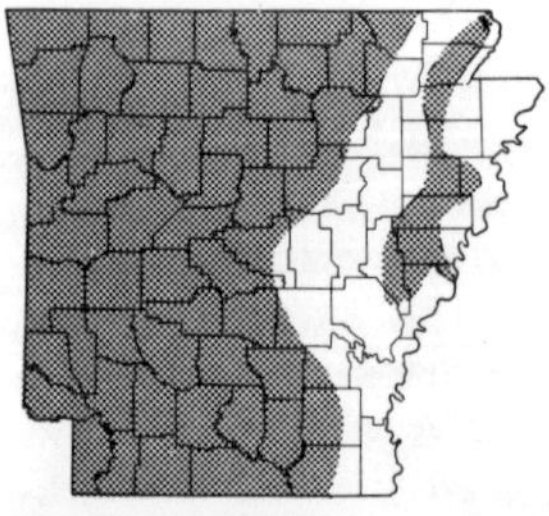

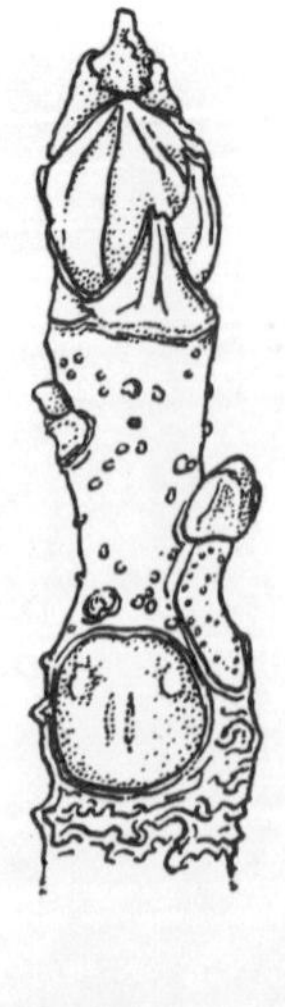

SHAGBARK HICKORY — *Carya ovata.* Terminal bud imbricated with dark, triangular, keeled, loose-fitting scales. The alternate lateral buds are much smaller. Leaf scar very widely obovate with three circular patches of vascular bundle scars. Twigs stout and dark. Old bark, light gray, shaggy and separating in long, thick plates, loose at the ends. Its fruit, with a sweet seed, is the most desirable of the state's "true hickories." Found on moist slopes and high floodplains with mixed hardwoods. *Juglandaceae.*

SUPERPOSED BUDS

Hickories-Walnuts : Trees. Leaf scars broad. Twigs stout. pp. 25-27

Virginia Willow : Delicate shrubs. Fruit tassels persist. Pith chambered. p. 27

Carolina Silverbell : Small tree. Twigs pale-brown, angular. Pith chambered. p. 28

Storax : Shrub. Leaf scar elongated, notched. p. 28

SINGLE LATERAL BUDS

Large-leaved Storax : Shrub. Twigs maroon. Buds buff. Wood green. p. 28

Sweetleaf : Small tree. Bud scales green, fleshy. Terminal leaves evergreen. p. 29

American Smoketree : Small tree. Buds bunched and purple at twig tip. p. 29

Common Alder : Shrub. Buds stalked, dark red. Fruits cone-like. p. 29

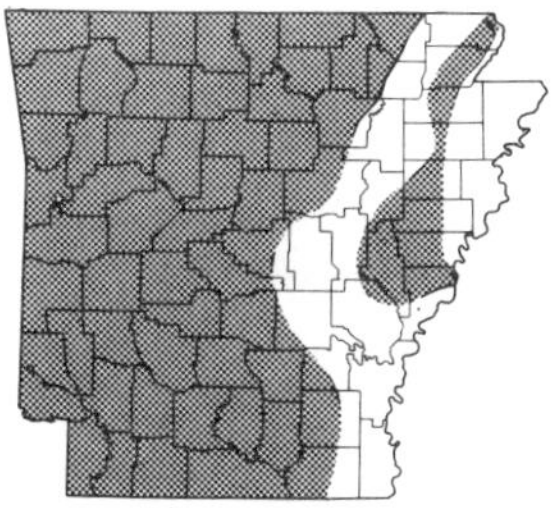

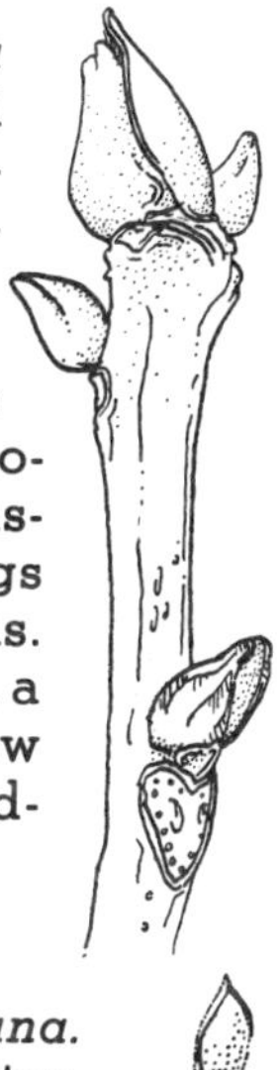

MOCKERNUT HICKORY — *Carya tomentosa.* Outer scales of terminal bud, few in number, thick and deciduous in autumn, exposing the yellowish tomentum of the inner scales. Lateral buds alternate, often superposed, and one-third the size of the terminal bud. Leaf scars widely obovate, notched at the top, and with vascular bundle scars scattered. The stout, hairy-to-smooth twigs are fragrant when bruised. Large trees with broad, open crowns. Limbs stout and smooth. Bark is thick, firm, smooth, and with a shallow diamond pattern caused by the intersection of narrow ridges. Common to well-drained slopes among mixed hardwoods. *Juglandaceae.*

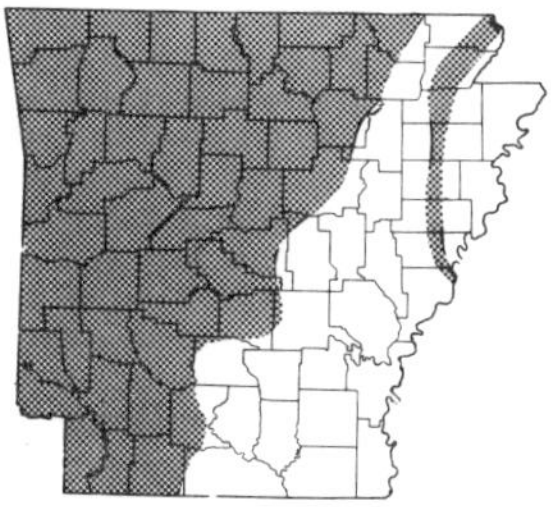

BLACK HICKORY — *Carya texana.* Terminal bud twice as large as the alternate buds. Bud scales and twigs with a rusty or red pubescence. Branches contorted and convoluted. Often found with blackjack oak, post oak, winged elm, and shortleaf pine, in dry habitats and savannas frequented by severe droughts and recurring fires. Old bark forms a diamond pattern with interlaced ridges of sometimes scaly bark. *Juglandaceae.*

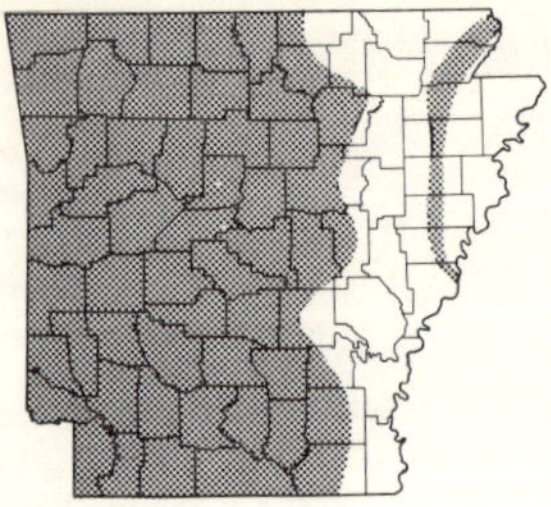

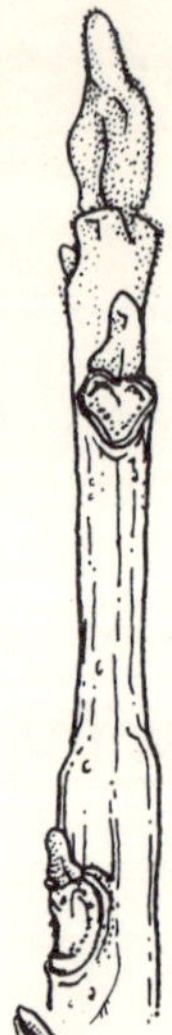

BITTERNUT HICKORY — *Carya cordiformis.* Terminal bud present. Lateral buds and leaf scars alternate. Buds valvate, scurfy, and bright yellow. Leaf scars broad with conspicuous vascular bundle scars. Twigs greenish-brown to gray-green. Bark of old trees is light gray, tight, smooth, and broken into faint, diamond patterns by shallow ridges and furrows. Found in well-drained bottom land and on low slopes. *Juglandaceae.*

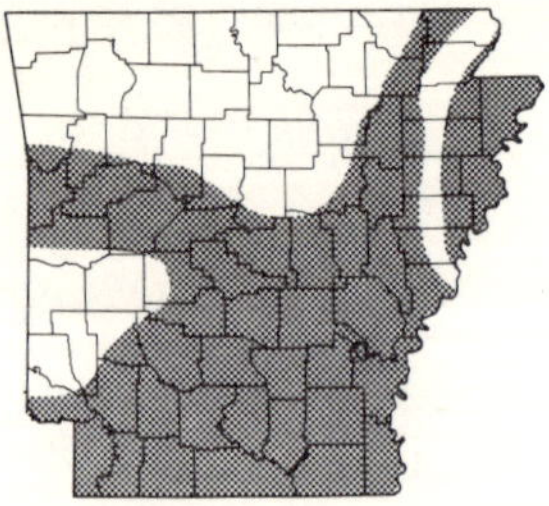

WATER HICKORY — *Carya aquatica.* The "bitter pecan" of bayous, delta rivers, and inundated lowlands. One of the three "pecan hickories," along with "sweet pecan" and nutmeg hickory. Terminal bud present. Lateral buds alternate, valvate, scurfy, and dull yellow. Twigs reddish-brown with orange lenticels. Stipular scars present. The general appearance of its limbing and bark resembles that of pecan. Its habitat preference distinguishes it; common only to soils subjected to persistent stream overflow. *Juglandaceae.*

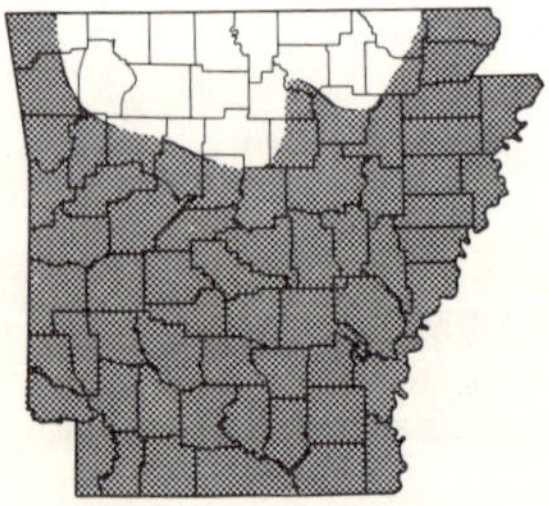

PECAN — *Carya illinoensis.* Terminal bud valvate and hairy. Lateral buds alternate above a broad indented three-lobed leaf scar. Vascular bundle scars numerous and scattered. Twigs hairy. Superposed flower buds common. Sought after and cultivated for its edible seed. Producing a broad, coarse crown in well-drained alluvial plains. *Juglandaceae.*

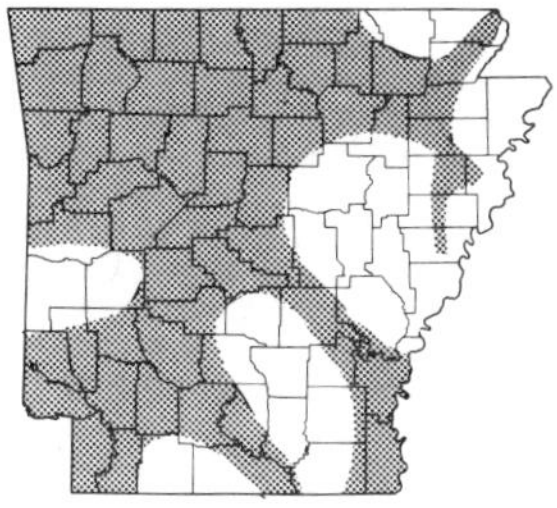

BLACK WALNUT — *Juglans nigra.* Terminal bud valvate with a few pale, pubescent scales. Lateral buds alternate, often superposed. Leaf scars elevated, obovate, notched at the top, and with three groups of vascular bundle scars, the lower group U-shaped. Stout twigs covered with rusty brown hairs. Pith bluff-colored and chambered. The thick, nearly black bark has deep, narrow furrows. Dull bark surfaces become polished brown when shaved with a knife. Globose fruits with black husks are to be found on the ground. Trees of moist, fertile soils, sought after for lumber and nut meats. *Juglandaceae.*

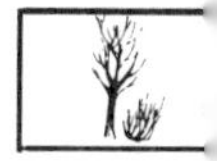

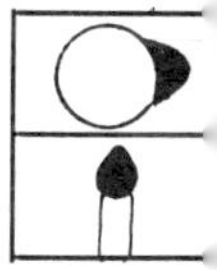

BUTTERNUT — *Juglans cinerea.* Terminal bud valvate with dull yellow and pubescent scales. Lateral buds alternate, often superposed. Leaf scar widely obovate with a prominent bottom lobe. Vascular bundle scars fused into three groups, those of the bottom lobe forming a "U". Twigs gray-green with dark-brown chambered pith. Leaf scar with velvety fringe forming a dense cushion below the bud. Ash-gray old bark is shallow and smooth. Husked fruits contain a deeply ridged shell and sweet, edible seed. This white walnut is a fast-growing, short-lived tree reaching the southwest limits of its range in Arkansas. *Juglandaceae.*

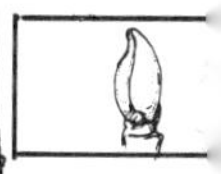

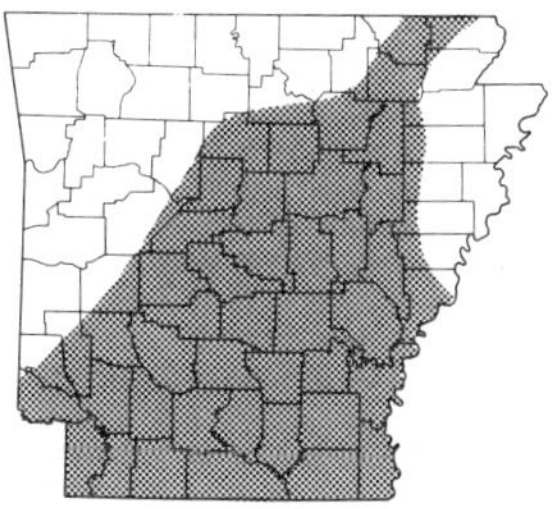

VIRGINIA WILLOW — *Itea virginica.* A short, delicate shrub of cypress swamps and sandy springheads. Flowers are produced on new wood. Fruits persist into winter as tassel-like dry racemes with two-parted fruits. The fruit capsules release many tiny, shiny, black seeds. Terminal buds valvate. Lateral buds alternate, often superposed. Leaf scar half elliptical with three groups of vascular bundle scars. Lower edge of leaf scar and lateral buds project outward and give the thin twig a rough texture. Bark of twigs and older stems glossy, mottled, mahogany, and green. Pith is chambered and granulated. *Saxifragaceae.*

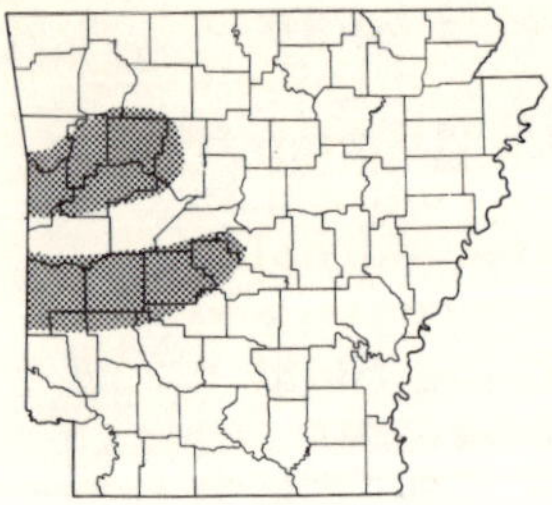

CAROLINA SILVERBELL — *Halesia tetraptera* var. *monticola.* Twigs pale brown, flattened and angular, swollen at the nodes because of the raised leaf scars. Leaf scars alternate, margins rimmed, vascular bundle scars grouped in a central U-shape. Terminal bud and lateral buds valvate with one or two sets of scales. Buds superposed; one large, one small, triangular and pointed. Old stems dark, streaked and finely shredding. Pith chambered. Find small, dry, winged fruits close to the stems of second year wood. A handsome small tree with a diffuse crown, of moist woods. *Styracaceae.*

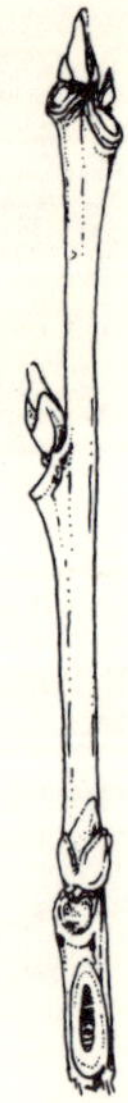

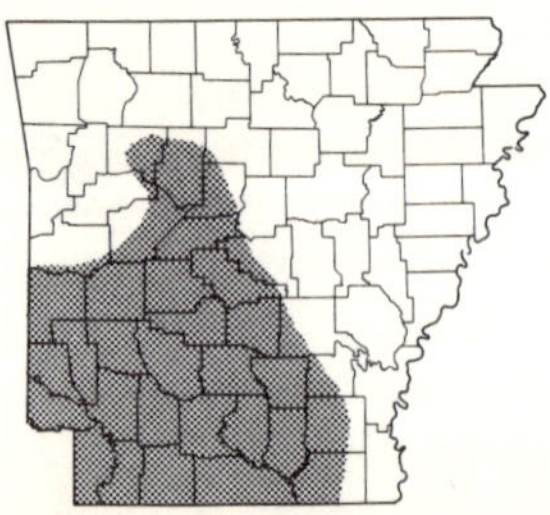

LARGE-LEAVED STORAX — *Styrax grandifolia.* Shrubs or slender small trees found on well-drained slopes or moist sandy ledges. Hirsute twigs maroon, blotched with gray, ending with a buff-colored, velvety, naked pseudo-terminal bud. Alternate lateral buds, small, globose, buff, velvety, located off center of an elliptical leaf scar. Vascular bundle scars of three groups: two, high on the sides and a central U-shaped patch. Wood green when cut then canary yellow with drying. Upland mixed deciduous and pine forests. *Styracaceae.*

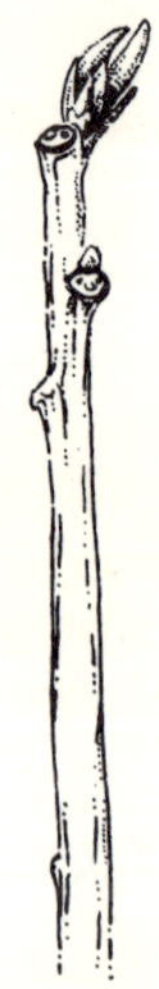

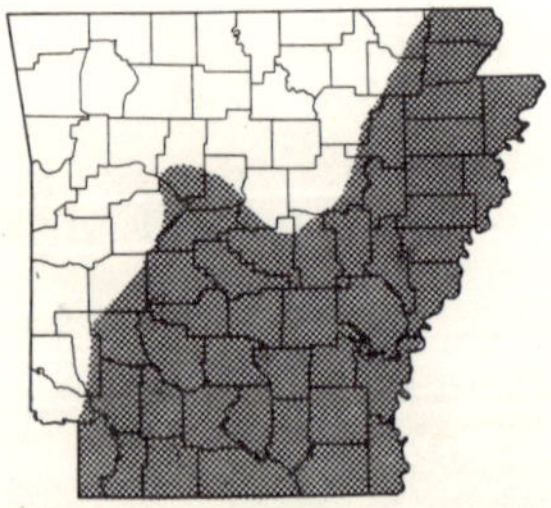

STORAX — *Styrax americana.* Shrubs found on overflow banks of streams in cypress swamps. Noticeable by smooth, black stems, streaked branches, and dull gray twigs with buff buds. Twigs are zigzagged and end with a terminal bud scar peg. Leaf scars are elevated, long-elliptical, deeply notched at the top. A crescent of denticular vascular bundles project from the central leaf scar. Three velvety lateral buds of unequal size are superposed above the leaf scar. Of wetlands forests. *Styracaceae.*

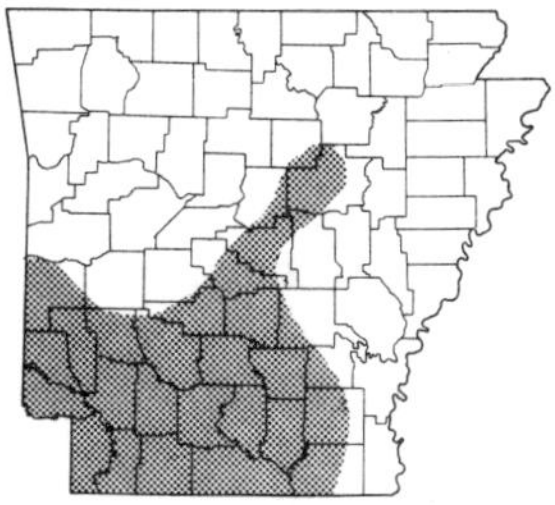

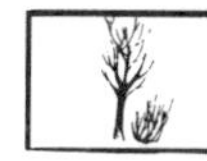

SWEETLEAF — *Symplocos tinctoria.* Terminal leaves are evergreen but others mostly deciduous. Terminal buds valvate with green fleshy scales, hairy at the tips. Twigs dulled by sparse, coarse, stiff hairs. Lateral buds alternate, valvate, dark, and small. The half-round leaf scars have a central ring of vascular bundle scars. The first lateral branch below the twig is perpendicular to the stem and vigorous. This condition leads to a widely forked and open crown. A small gray tree of sandy woods, swamps, and bottomland. *Symplocaceae.*

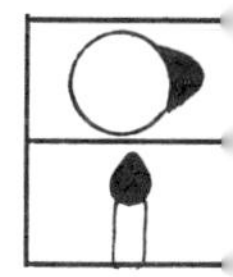

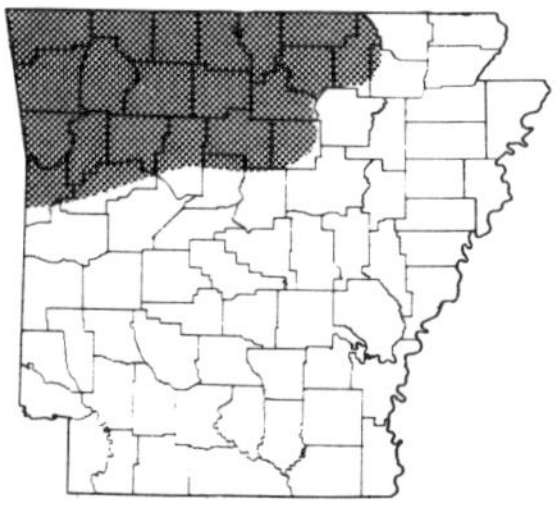

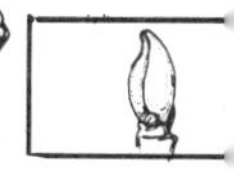

AMERICAN SMOKE TREE — *Cotinus obovatus.* The terminal bud and several alternate, lateral buds are bunched at the twig tip. Buds numerous and valvate. Buds and leaf scar borders, deep purple. Leaf scars are raised but the surface is sunken below the level of three vascular bundle scars. Twigs smooth, brown, and aromatic. Branches and old stems gray and dull with a brown pith. A small tree with a wide dense crown. Isolated populations or single specimens on bluffs and ledges of the Ozarks. *Anacardiaceae.*

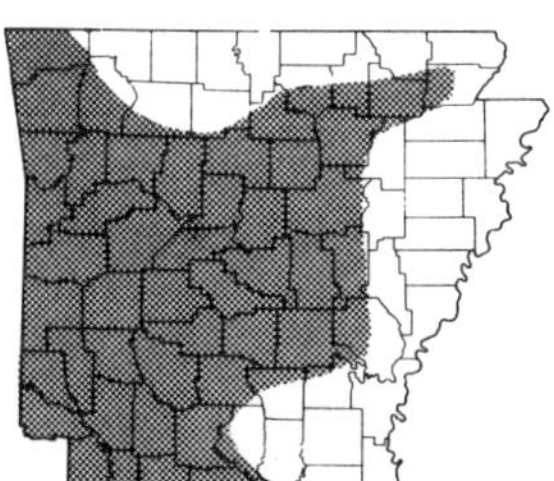

COMMON ALDER — *Alnus serrulata.* Terminal bud present. Lateral buds alternate, valvate, stalked, and dark red. A short stipular scar is present on each side of the leaf scar. The female catkins form woody, cone-like, mature fruit. Male catkins are elongated. A hard, coarse shrub found commonly along stream banks and discovered easily by its fruits. *Betulaceae.*

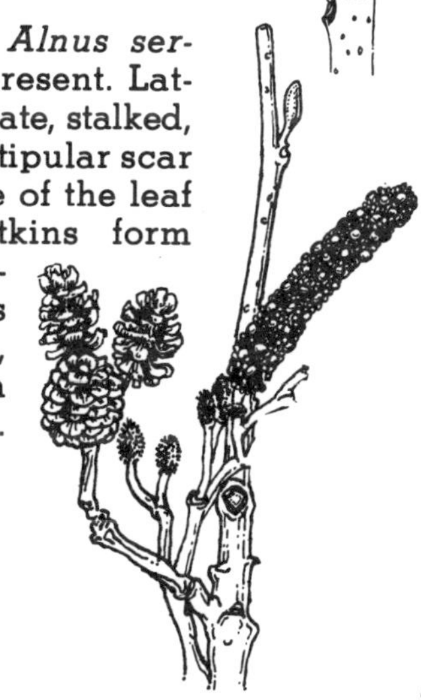

Witch Hazel : Terminal bud long, flattened, velvety, tan to yellow. p. 30
Paw Paw : Terminal bud long, flattened, velvety, dark brown. p. 30
Carolina Buckthorn : Terminal bud long, pointed, downy. Bark has a rank odor. p. 31

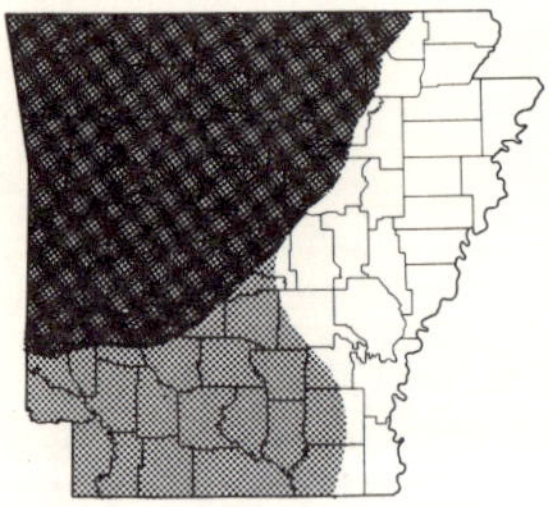

WITCH HAZEL — *Hamamelis vernalis.* Terminal bud long and flattened. Buds alternate, naked, velvety, tan to yellow. Flowers with yellow strap-like petals turning reddish with age. A coarse, large shrub suckers freely from the base. Flowers in late winter and early spring. On gravel bars and in rocky streambeds of swift upland watercourses. *H. virginiana,* the autumnal-flowering witch hazel, does not sucker, and occurs only on streambanks and floodplains. The species resemble one another closely, and both yield bark with astringent properties. The map shows both ranges. *H. virginiana* overlays *H. vernalis* and extends southward into the Gulf Coastal Plains. *Hamamelidaceae.*

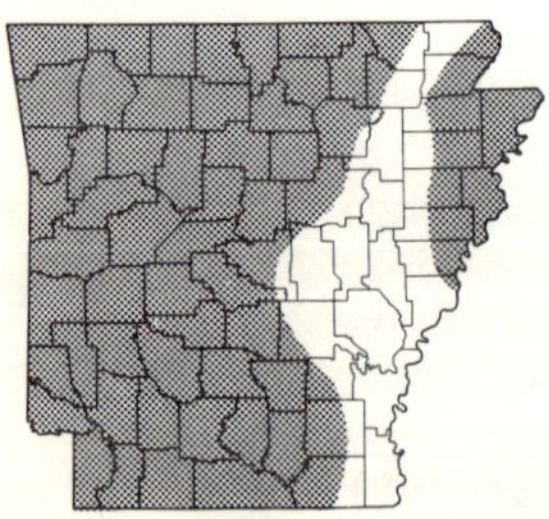

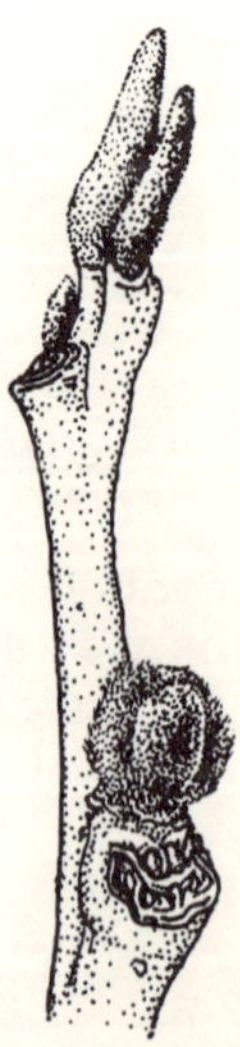

PAWPAW — *Asimina triloba.* Terminal bud long and flattened. Lateral buds and rough leaf scars alternate. Buds naked, velvety, dark brown. Flower buds lateral and globose. Bruised bark of twigs gives off a distinct oily odor. A small tree of moist, shaded habitats. *Annonaceae.*

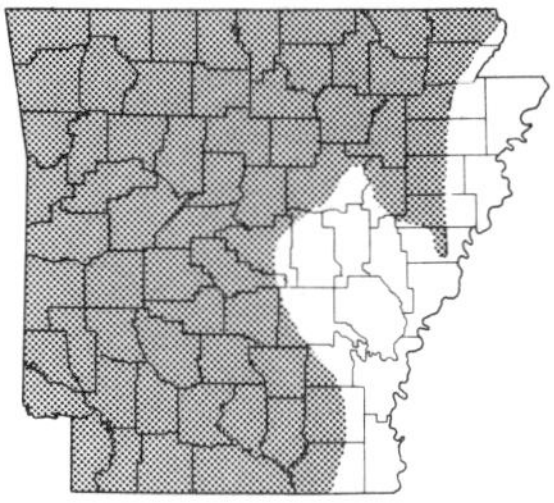

CAROLINA BUCKTHORN — *Rhamnus caroliniana.* Naked terminal bud, long, pointed, dull, downy, with short hair. Round lateral buds are comparatively tiny, of the same color, and hairy as well. Alternate leaf scars protrude at the bottom side and are followed by shallow ridges down the twig. Three vascular bundle scars appear near the center of the leaf scar. Torn bark yields a rank odor. Shiny, black, fleshy fruits last into winter. A tall shrub forming thickets in moist woods. *Rhamnaceae.*

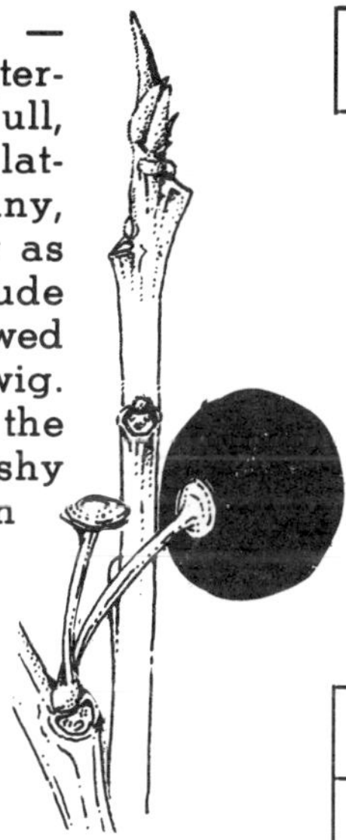

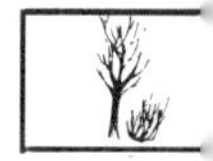

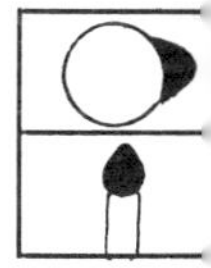

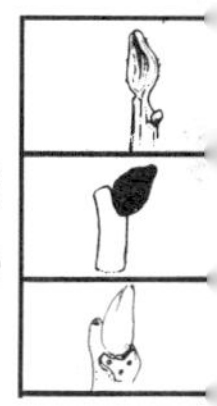

TREES/SHRUBS Pseudo-Terminal **Alternate Buds**

Conicate

Sycamore : Twigs zig-zagged, golden. Buds dark-red, resinous. p. 31
Leatherwood : Twigs jointed, rubbery. Buds gray, hairy. p. 32
Willows : Twigs yellow or green, brittle. Die-back common. Stipular scars present. pp. 32, 33

SYCAMORE — *Platanus occidentalis* Terminal bud lacking. Lateral buds alternate, dark-red, resinous, and covered with a cone-like outer scale. The narrow leaf scar nearly encircles the bud. Twigs strongly zig-zagged. Branches chalk-white with a smooth bark peeling away in long thin scales. Twigs golden. Young branches gray-green. Riverbanks, gravelbars, streamsides, and floodplains. *Platanaceae.*

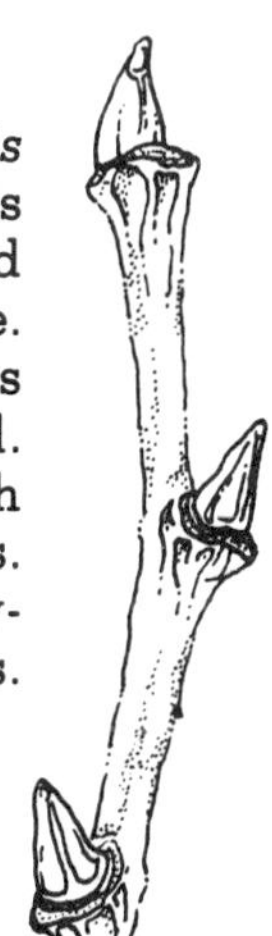

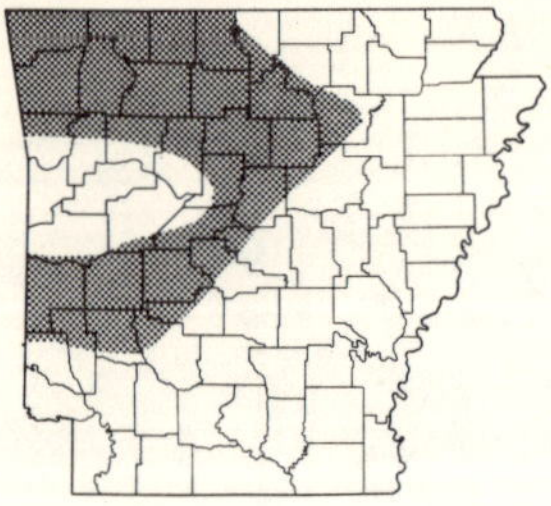

LEATHERWOOD — *Dirca palustris.* A shrub with extremely lightweight wood and supple branches. The rubbery bounce of a teased shrub is a unique and reliable characteristic. Peeled bark has the feel and strength of leather. The bulbous twig tip consists of a terminal bud scar peg and a globose, gray, tomentose psuedoterminal bud nearly surrounded by a leaf scar. Insertions of branches are from the center of leaf scars and give an articulation to the growth form. The forked pattern of branching is established by the growth of the first alternate, lateral bud below the psuedoterminal bud. Stems are practically pithless. Upland forests of ravines, north slopes, and well drained floodplains of the Boston and Ouachita Mountains are its common habitats. *Thymelaeaceae.*

BLACK WILLOW — *Salix nigra.* Twigs long, unbranched, round, drooping, brittle at the base, but tough and flexible above. Tip die-back most common or with a pseudo-terminal bud. Buds alternate, covered with a single cap-like scale. The crescent leaf scars have stipular scars. Shrubby young trees grow into tall, multi-trunked trees with flaky, dark-brown bark. Growing in wet soils in zones of fluctuating water levels. *Salicaceae.*

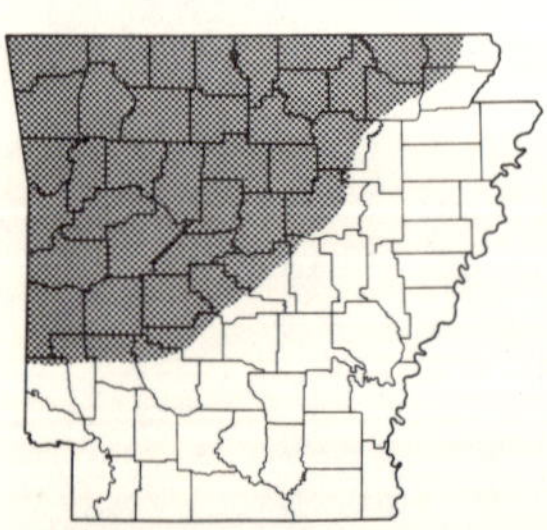

WARD'S WILLOW — *Salix caroliniana.* Tip die-back common. Cone-buds are alternate, pubescent, pointed, and with the tip appressed. The leaf scars are elevated, rimmed, and bluntly obtriangulate. Three vascular bundle scars project above the sunken leaf scar surface. Twigs brittle at the base. Coarse shrubs, often no more than three meters tall, with gray, deeply checkered bark. Found along stream ledges, banks, gravel bars, intermittent stream beds, and water courses. *Salicaceae.*

BALL WILLOW — *Salix matsudana.* Called parasol willow by some. A conspicuous winter tree of the open, flat, delta lands and south Arkansas because of its yellow-green, full cone- or umbrella-shaped crown of fine textured branches. Commonly planted around house yards, headquarters, and parks. Psuedo-terminal bud accompanied by a terminal bud scar peg. Lateral buds alternate, enclosed with a single fleshy bud scale that wraps around to a flat inside face. Bud scales ribbed on the side. Leaf scar a thin crescent with stipular scars at the ends. Propagated best by autumn cuttings. Probably the "umbraculifera" cultivar of *S. matsudana,* the Pekin willow. *Salicaceae.*

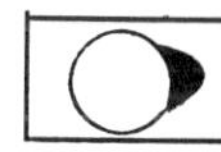

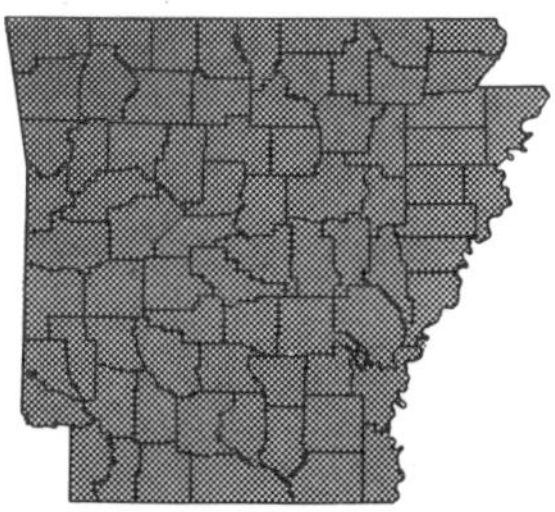

WEEPING WILLOW — *Salix babylonica.* "Willow whips" have shiny, smooth, golden-yellow twigs and buds. Pseudo-terminal bud hooks to one side over a terminal bud scar. Buds are alternate and covered with a single bud scale. Lateral buds are long-pointed with the tip appressed and blended with the twig profile. Raised leaf scars are narrow, lobed crescents with three vascular bundle scars, the middle one projecting. Stipular scars present as single dots at the end of each leaf scar. Introduced and cultivated near water for its vigor and pendulous, diffuse growth form. *Salicaceae.*

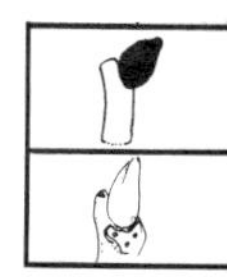

STIPULAR SCARS

Tulip Tree and Magnolias : Stipular scar encircles twig. pp. 34, 35
Ozark Chinquapin : Stipular scar small, restricted. Buds red, black-tipped. p. 36
Chinese Tallow Tree : Stipular scars warty. Twigs striated. p. 36
Possum Haw : Stipules withered or scar faint. Buds purple. Berries red. p. 36

NO STIPULAR SCARS

BROAD LEAF SCARS

Tree-of-Heaven : Twigs stout, downy. p. 37
Chinaberry : Twigs stout, smooth. p. 37
Kentucky Coffee Tree : Twigs stout, mottled. Superposed buds. p. 37
Soapberry : Twigs downy with yellow hairs. Superposed buds. p. 38
Mimosa : Twigs glossy green. Superposed buds. p. 38
Golden Rain Tree : Twigs brown, warty. Leaf scar outlined in black. p. 38
Indigo Bush : Y-shaped leaf scar. Superposed buds. p. 39

SMALL OR CIRCULAR LEAF SCARS

Sumacs : Die-back common. Pithy stems. pp. 39, 40
Persimmon : Bud scales persist at base of branches. p. 40
Male-Berry : Buds red. Twigs dull gray. p. 40
Staggerbush : Buds red, long-pointed, curved. Twigs dull brown. p. 41

TULIP TREE — *Liriodendron tulipifera*. Terminal bud is flattened, glabrous, and valvate. Lateral buds are tiny, globose, and alternate. Leaf scars are widely elliptical with stipular scars encircling the twig. Twigs are lustrous and dark purple. The bark of older stems is thin, tight, shallowly furrowed and ash gray. Large erect, cone-like aggregates of winged, four-angled fruits appear high in the crown. A fast growing pioneer tree of disturbed soils on Crowley's Ridge. Widely cultivated as an ornamental and escaping occasionally. *Magnoliaceae*

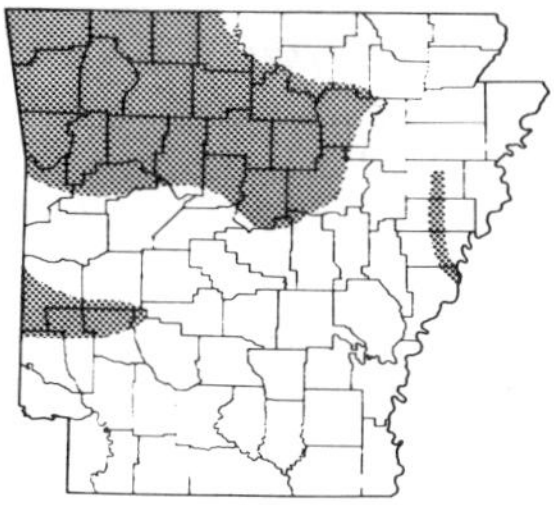

CUCUMBER TREE — *Magnolia acuminata.* Pseudo-terminal bud is large with valvate scales covered with fine, white hairs. Lateral buds are alternate, valvate. Leaf scars are narrow crescents with vascular bundle scars lined up in a row. Stipular scars encircle the twig. Twigs are stout and lustrous. Brown, thin, bark of old trees is shallow and has narrow, scaley ridges. Cone-like fruits can be found on the ground. This large tree is found in moist sites of upland hardwoods. Moves into areas as a result of rock slides, mud slides, logging roads, stream meandering, and windfalls. *Magnoliaceae.*

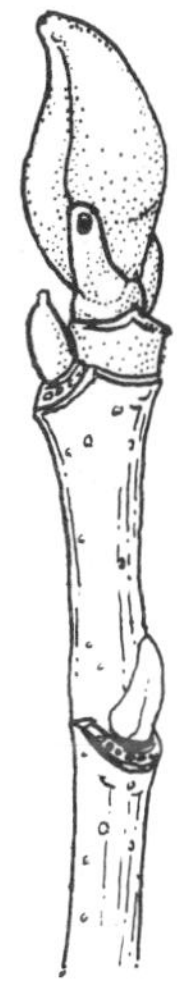

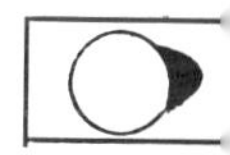

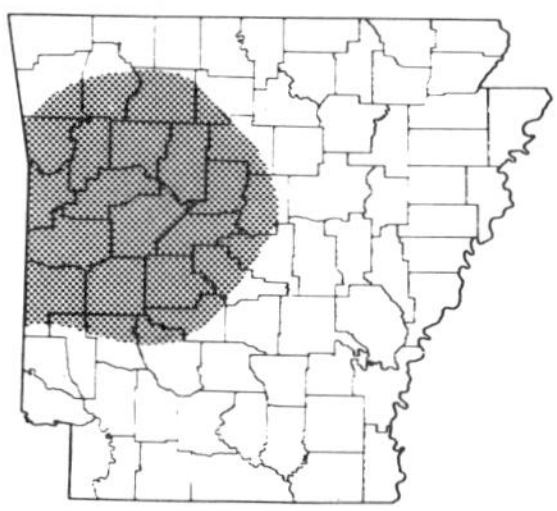

UMBRELLA MAGNOLIA — *Magnolia tripetala.* Pseudo-terminal flower bud is large, round, long-pointed, valvate, and covered with a waxy bloom. Lateral buds are alternate. Leaf scars are circular to half-round, broad, with stipule scars. Twigs are bright reddish-brown, lustrous, with scattered pale lenticels. Bark is light gray, smooth, and thick. Several stems often grow from near the base of the trunk forming a large spreading bush. With beech and sugar maple in moist organic soils along mountain streams and deep coves. *Magnoliaceae.*

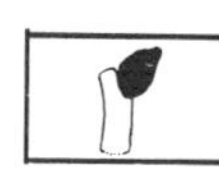

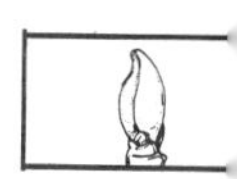

SOULANGE'S MAGNOLIA — *Magnolia soulangiana.* Large pseudo-terminal flower bud is valvate; scales have a coarse, gray pubescence. Lateral buds and leaf scars are alternate. Vascular bundle scars are scattered. Stipular scars are present. Flower and fruit scars are common in the forks of branches. A large shrub with smooth, ash-gray bark. Introduced, cultivated, and prone to pre-seasonal flowering in warm, late winters. *Magnoliaceae.*

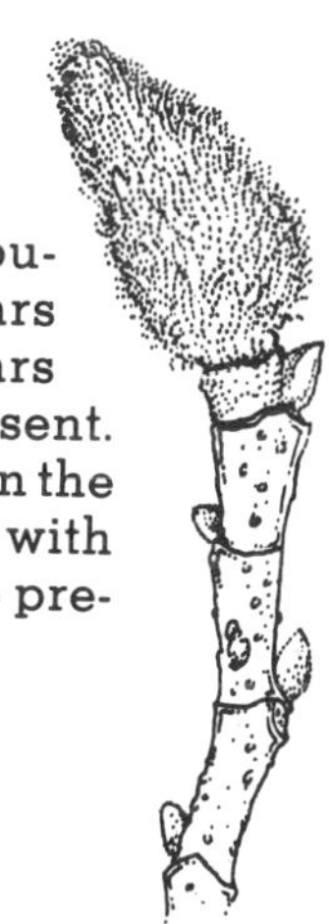

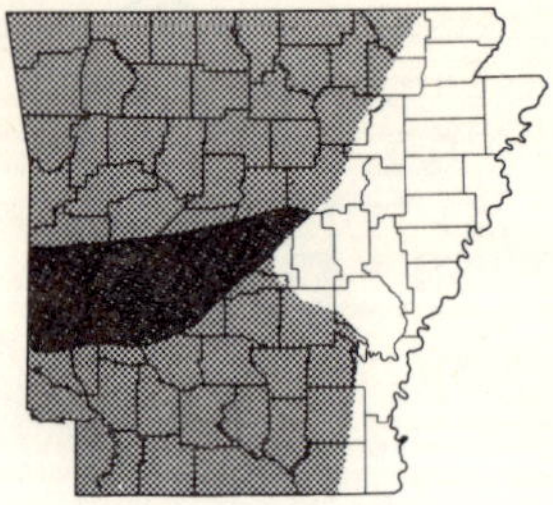

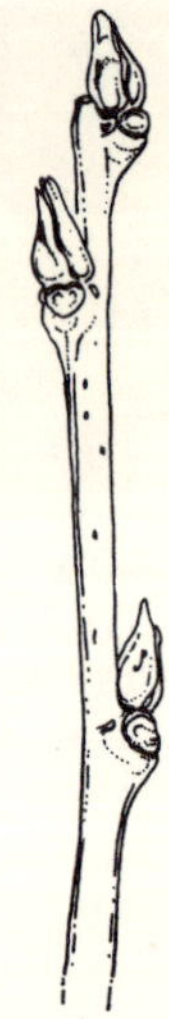

OZARK CHINQUAPIN — *Castanea pumila* var. *ozarkensis.* Twigs lustrous gray-brown, flecked with white lenticels. Pseudo-terminal bud has a terminal bud scar at the twig tip. Buds alternate, oblique, and valvate with black-tipped, reddish-brown scales. Leaf scars raised, half-round, with scattered vascular bundle scars and distinct stipular scars. Trees commonly killed back by disease and sprouting from the root base. This variety of the chestnut is of the north half of the state; *C. pumila* var *pumila* is to the south. Find spiny burs as remnants of the fruit. *Fagaceae.*

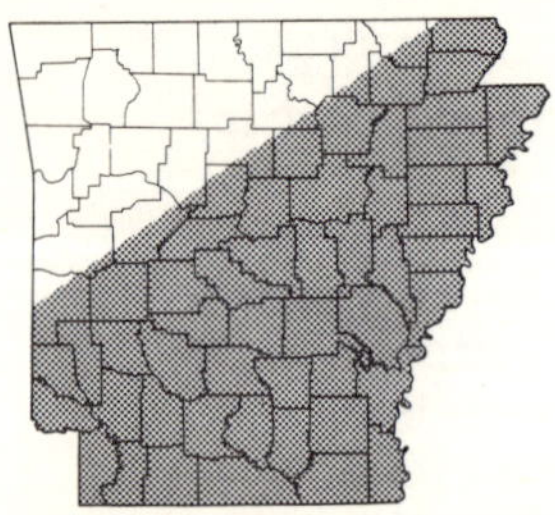

CHINESE TALLOW TREE — *Sapium sebiferum.* The limbs and branches of the broad crown are repeatedly forked. Old bark is gray with light brown fissures. Valvate lateral buds are alternate, squat, and triangular. Half round leaf scars showing three vascular bundle scars are flanked by warty stipule scars. Twigs faintly striated by shallow ridges. The white, fatty seed-coverings are used for candle and soap making. An introduced, cultivated, and escaping tree subject to top-kill in severe winters, or otherwise, twig die-back. Seed-foraging birds account for much of its natural spread. *Euphorbiaceae.*

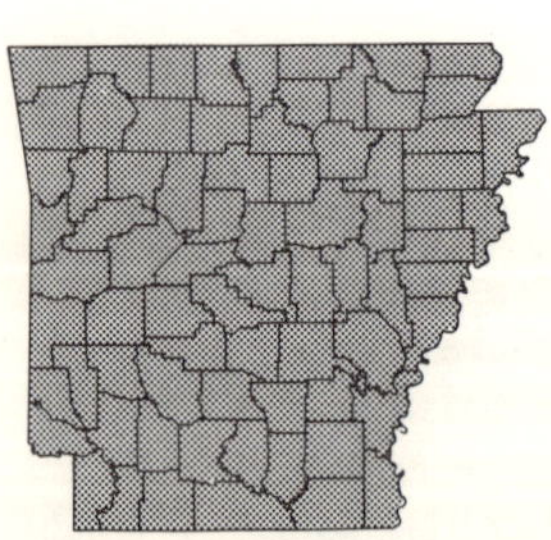

POSSUM HAW — *Ilex decidua.* A small, deciduous holly tree with a loose, wide-spreading crown. Small, shiny, red berries are produced annually. Berries have four to five seeds embedded in a sticky yellow matrix. Repeated seed crops on the same branch produce fruit spurs and dwarf branches. Twigs gray-green, pubescent, and flaky. Psuedo-terminal buds valvate, sharp-pointed, and deep purple. Twigs blunt by terminal bud scar and leaf scar. Lateral buds alternate, sharp, and valvate. Leaf scars raised, half-round, with two central vascular bundle scars often appearing as one. Short, sharp, withered stipules or faint stipular scars attend the leaf scar. These small trees are found along fencerows, in thickets, and in low woods. *Aquifoliaceae.*

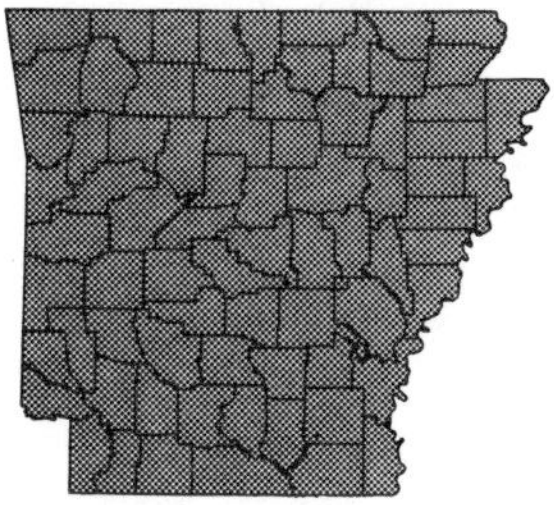

TREE OF HEAVEN — *Ailanthus altissima.* Twigs stout and downy with short, soft hairs. Twig tip coarse with projecting leaf scars and a broad terminal bud scar. Leaf scars broadly heart-shaped with a line of vascular bundle scars following the margin. The low, blunt valvate lateral buds sit in the leaf scar notch. Bruised bark emits a disagreeable odor. Brittle twigs show a brown pith. Introduced, cultivated around dwellings and spreads from root sprouts in rural and urban areas. *Simaroubaceae.*

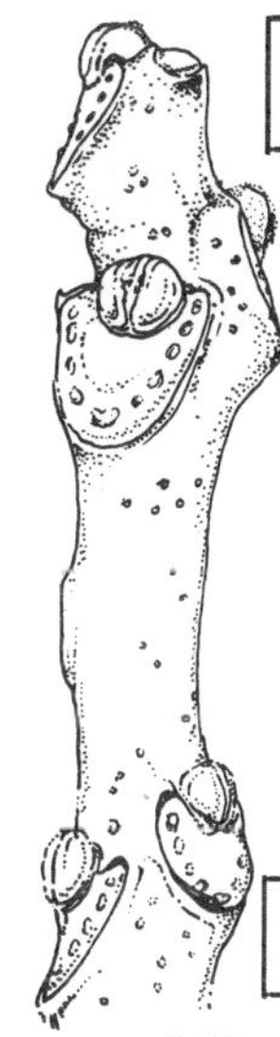

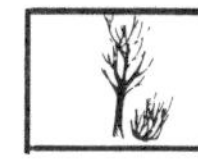

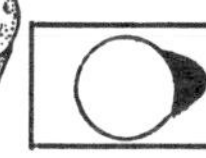

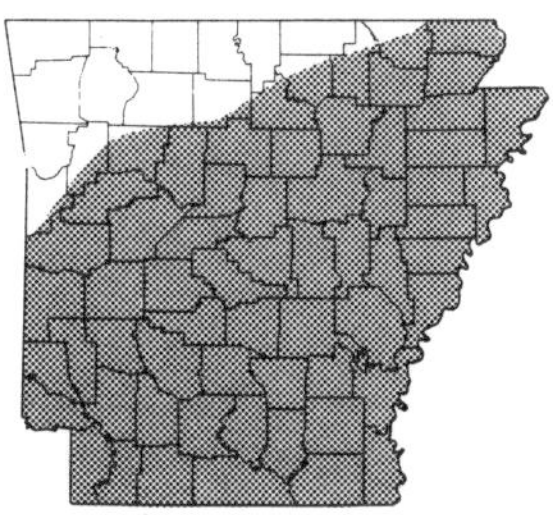

CHINABERRY — *Melia azedarach.* Dark, stout, glabrous twigs have hairy valvate buds. Leaf scars raised, broad, three-lobed with three groups of vascular bundle scars. A vigorous, small tree, usually with an abundance of pale yellow, pulpy fruits in loose panicles lasting into winter. Introduced and cultivated. *Meliaceae.*

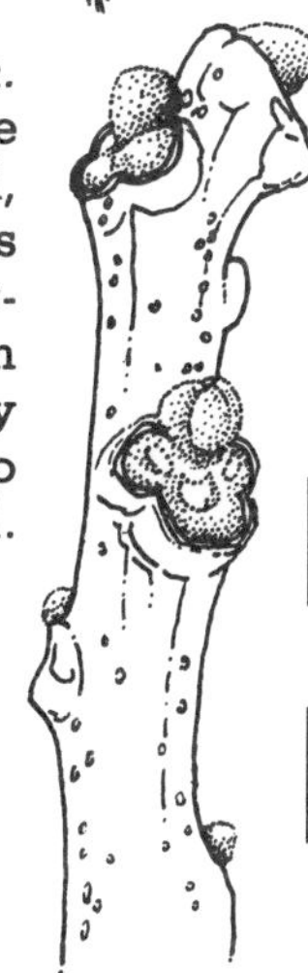

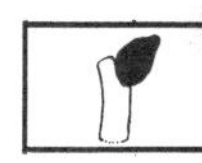

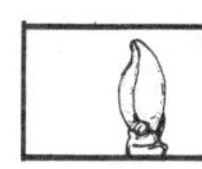

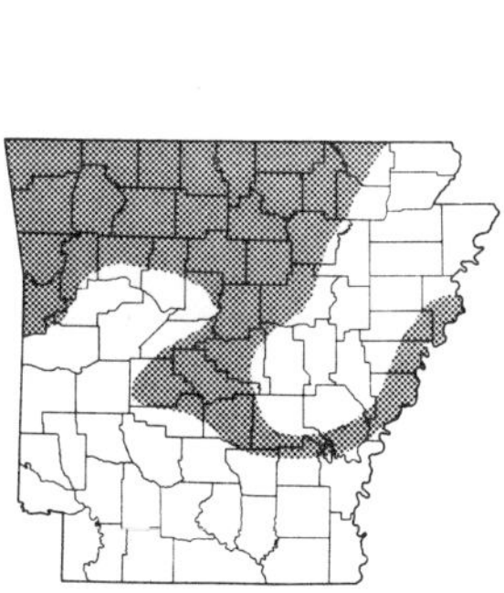

KENTUCKY COFFEE TREE—*Gymnocladus dioicus.* Twigs stout and straight with strong contours. Lateral buds superposed and each embedded in a hairy crater. Leaf scars broad, three-lobed, with three groups of vascular bundle scars. Sharp, fluted bark first shows on young, vigorous trees and is a persistent characteristic to maturity. A rare tree in natural habitats but propagated commonly by seedlings grown from the large beans. *Leguminosae.*

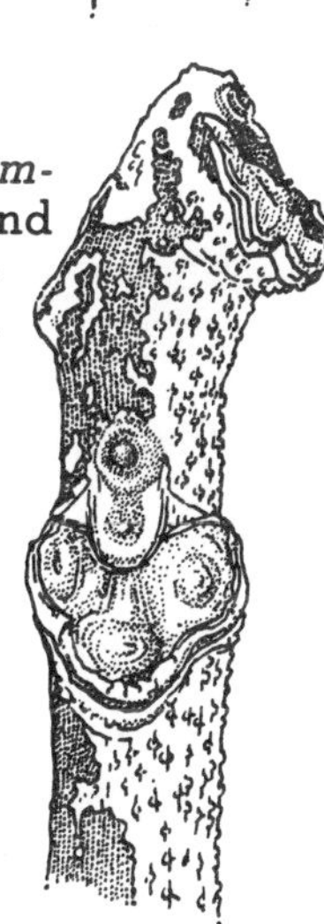

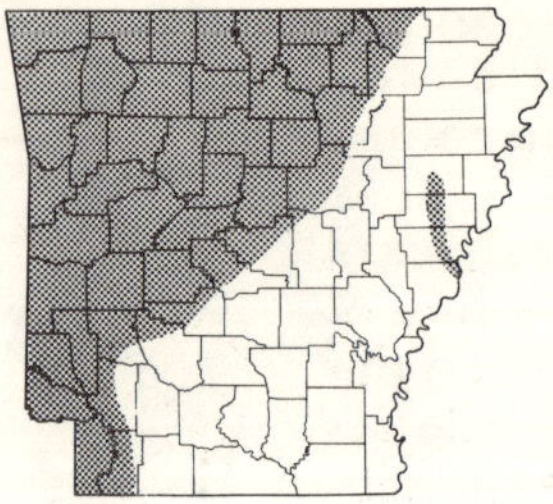

SOAPBERRY — *Sapindus drummondii.* A slender, medium sized tree with fluted bark and dull, light-gray branches. Twigs downy, pale gray with alternate buds. Leaf scars broad, ovate or heartshaped, with three patches of vascular bundle scars. Buds, small, globose, valvate, and downy with pale yellow hairs. Superposed buds common on vigorous sprouts and seedlings. Terminal bud absent. Isolated populations are found at the base of limestone cliffs propagating by seed into sandy banks of nearby streams. *Sapindaceae.*

MIMOSA — *Albizia julibrissin.* Twigs glossy, dark-green, terminated by a bud-scar post or by remnants of fruit clusters. Lateral buds imbricate and superposed above broad, shield-shaped leaf scars. A vigorous, small tree with a widely spreading crown above multiple or divergent trunks. An old world species that is widely cultivated. *Leguminosae.*

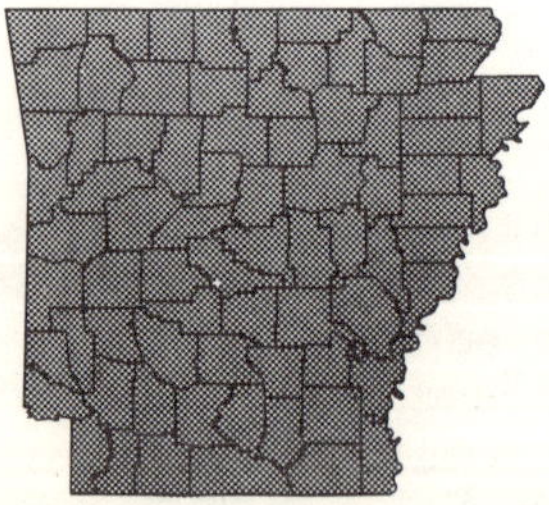

GOLDEN RAIN TREE — *Koelreuteria paniculata.* Buds brown, short, triangular, valvate and laterally compressed. Leaf scars shield-shaped and fringed in black. Lenticels warty. A small, cultivated tree with loose panicles of papery fruits lasting into winter. *Sapindaceae.*

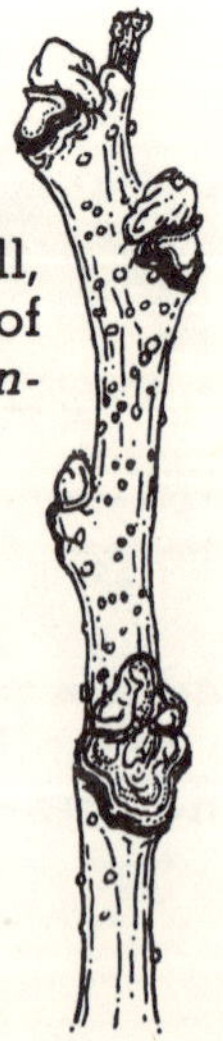

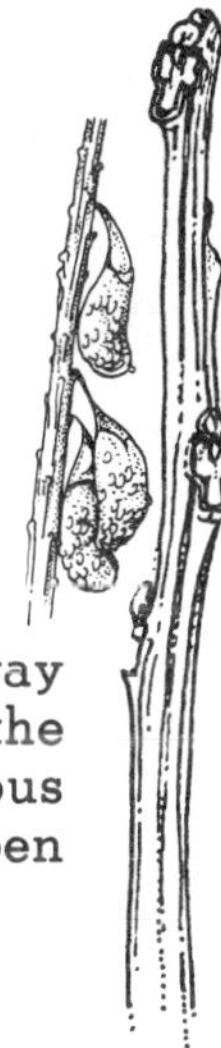

INDIGO BUSH — *Amorpha fruticosa.* Twigs blunted by a flat, oblique terminal bud scar. Lateral buds alternate, short, conical, round to diamond-shaped in outline, and tawny, with coarse, brown hairs. Buds superposed with the smaller and lower bud in the notch of a broadly Y-shaped leaf scar. The narrow, lower lobe of the leaf scar has a pouting lip. Three distinct patches of vascular bundle scars are circular or crescent shaped. Twigs are faintly striated. Limbs are speckled with brown lenticels on a smooth, dull, gray bark. Branches arise at a low angle forming a slender "V" with the limb. Mature fruits persist into winter and show conspicuous glandular dots. This small, slender tree is seen on low slopes, open river banks, and in open floodplains. *Leguminosae.*

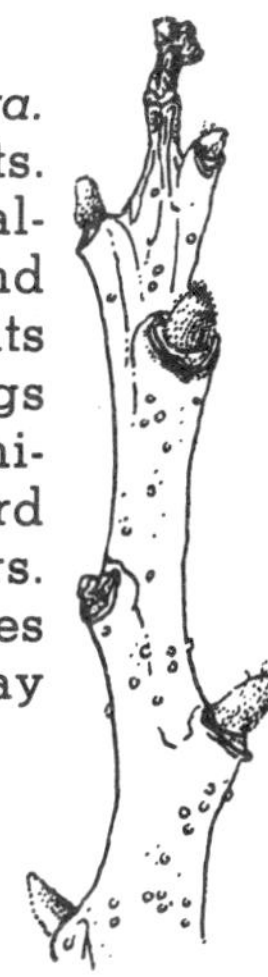

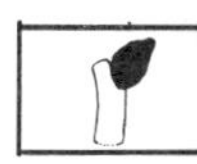

SMOOTH SUMAC — *Rhus glabra.* Die-back at tip or with terminal fruits. Pseudoterminal and lateral buds valvate. Lateral buds alternate, white and hairy. Leaf scars narrow crescents nearly surround the lateral buds. Twigs stout, glabrous, and pithy. Erect terminal fruit clusters have many small hard fruits covered with crimson hairs. Shrubs with sparse, forked branching. Producing wide clones by root sprouts in cleared or disturbed forest land, highway margins, and old fields. *Anacardiaceae.*

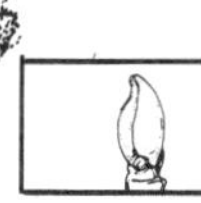

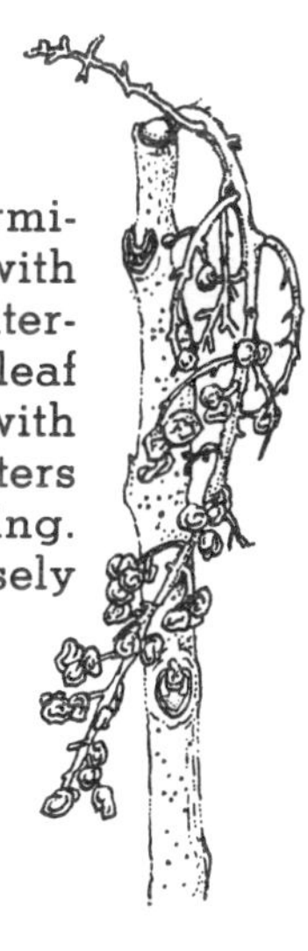

WINGED SUMAC — *Rhus copallina.* Twigs with die-back or terminal fruit clusters. Pseudoterminal and lateral buds are valvate with hairy scales. Lateral buds are alternate and inside a crescent-shaped leaf scar. Twigs stout, wooly, and with greenish pith. Terminal fruit clusters are dark red, withered, and drooping. A shrub of dry hillsides and open fields with a diffuse sparsely branched crown. *Anacardiaceae.*

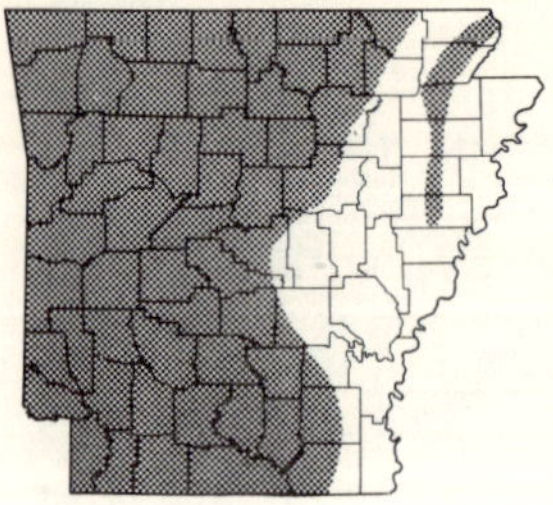

AROMATIC SUMAC — *Rhus aromatica.* Leaf scars elevated and round with yellow, hairy, embedded, lateral buds. The brown, smooth twigs are fragrant when bruised. The elongate and scaly flower-bud clusters are stalked and arise from the lateral buds. This knee- to chest-high straggling shrub is associated with woods of dry slopes and fire glades. *Anacardiaceae.*

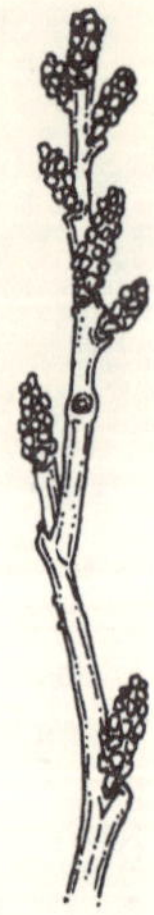

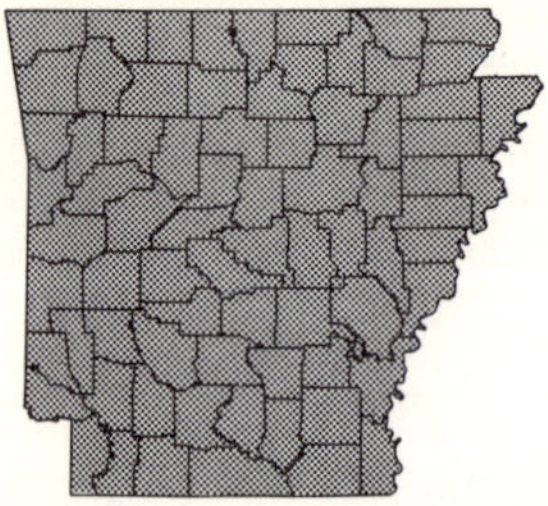

PERSIMMON — *Diospyros virginiana.* Pseudo-terminal bud present. Lateral buds alternate, valvate, dark, and each is subtended by a half-round leaf scar. Vascular bundle scars form an elliptical pattern in the center of the leaf scar. Bud scales are persistent and apparent at the base of young branches. Twigs hairy. Branch pattern shows the regular domination of terminal growth by two lateral branches. Black bark is divided into blocks. Numerous flat, shiny, brown seeds can be found on the ground under the canopy. Longitudinal sections of the seeds reveal a white embryo, the "knife" and "spoon" of folklore weather prediction. A highly productive pioneer tree of oldfields and disturbed forests. Mature trees have a high commercial value. *Ebenaceae.*

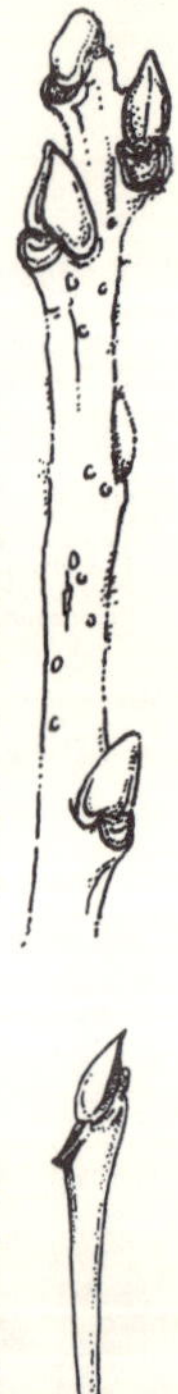

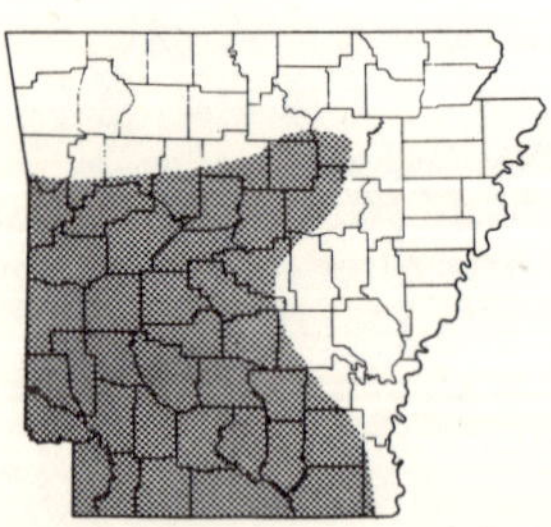

MALE-BERRY — *Lyonia ligustrina.* A dense-crowned shrub with dull gray shredding bark, red buds, and twigs. Twigs terminated with a bud scar and a psuedo-terminal bud. Buds alternate and valvate. Leaf scar half-elliptical with a central, U-shaped group of vascular bundle scars. Found in sandy streamsides, springheads, and lake margins. *Ericaceae.*

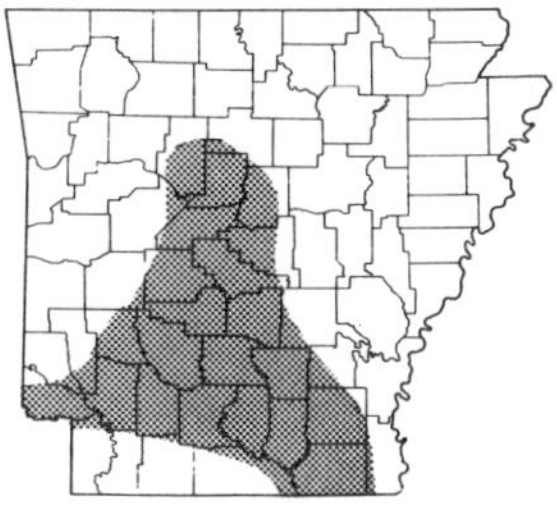

STAGGERBUSH — *Lyonia mariana.* The alternate buds are bright red, shiny, valvate, curved, long-pointed and on a dull brown stem. Leaf scars are half-round. Five-angled, open fruits persist into winter. Reported to be poisonous to young herbivores. A shrub of open, sandy woods, and sandy streambanks. *Ericaceae.*

TREES/SHRUBS Pseudo-Terminal **Alternate Buds** Imbricate

- **End Bud Larger Than Side Buds**

SHRUBS

Azaleas, Huckleberries, Blueberries : Leaf scar inconspicuous. Hairy flower buds common. pp. 43-45

Black Alder : Stipules are withered tags. Twigs ash-gray. Buds straw-colored. p. 45

Spicebush : Twigs and buds glossly green, aromatic. p. 45

Rose-of-Sharon : Stipular scars present. Fruits five-celled. p. 46

TREES

Ginkgo : Dwarf branches. Finely shredding bark. p. 46

River Birch : Twigs wiry. Buds long-pointed, curved. Bark papery. p. 46

Gums : Leaf scars with three conspicuous bundle scars. p. 47

Basswood : Twigs gray-green, smooth. Buds lop-sided, orange or red. p. 48

Sassafras : Twigs green, aromatic. Buds yellowish-green. p. 48

Wild Black Cherry : Twigs red-brown, pungent. Buds reddish. Stipular scars present. p. 48

• **End Bud Equal In Size To Side Buds**

STRAIGHT TWIGS

Bois D'Arc : Twigs with milky sap. Buds superposed. Fruits large. Bark and roots orange. p. 49

Red Mulberry : Twigs with milky sap. Buds brown. Root bark orange. p. 49

Redbud : Twigs dark, shiny. Flower buds globose, superposed, and collateral. p. 49

ZIG-ZAGGED TWIGS

Snow Wreath : Shrub. Twigs twisted, bowed, and later ridged. p. 50

Hazelnut : Shrub. Twig internodes curved. Fruit with a papery covering. p. 50

Buckthorn : Small tree. Twigs green with inconspicuous buds and stipular scales. p. 50

Wild Plum : Small tree. Twigs dark, glossy. Leaf scar indented. p. 51

Hop Hornbeam : Small tree. Twigs finely shredded. Bud scale striated. p. 51

American Hornbeam : Small tree. Buds downy-tipped. Trunk sinewy. p. 51

Planer Tree : Small tree. Buds hairy, superposed. p. 52

Red Elm : Tree. Twigs and buds hairy, reddish. Layered bark cinnamon. p. 52

American Elm : Twigs and buds smooth. Bark with alternating layers of light and dark. p. 52

Winged Elm : Tree. Corky wings on twigs and branches. Uplands. p. 53

Cedar Elm : Tree. Twigs and buds dull gray. Lowlands. p. 53

Siberian Elm : Tree. Twigs gray. Buds purple. p. 53

• **Die-Back At Twig Tip**

Sugarberry : Tree. Buds triangular, appressed. p. 54

Hackberry : Tree. Buds long-conical. Old branches with "witches brooms". p. 54

New Jersey Tea : Shrub. Twigs rough, dull. Disk-like fruits persist. p.54

Sea-Myrtle : Shrub. Twigs supple, angled, aromatic. Daisy-like fruits. p. 55

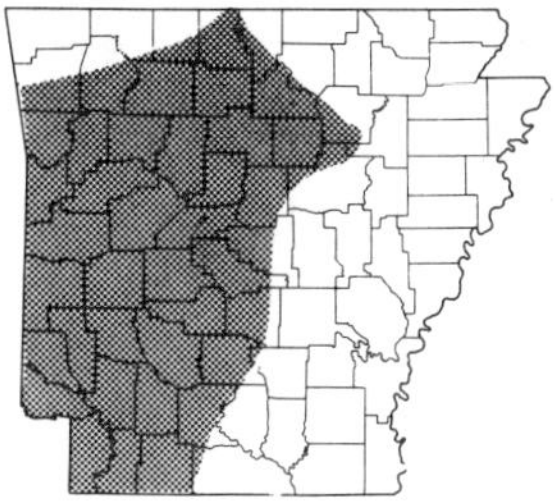

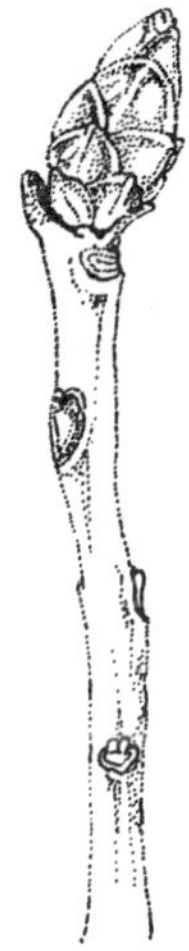

ELECTION PINK — *Rhododendron prinophyllum.* Twigs, vegetative buds, and flower buds with a common dull, light-brown color and short pubescence. Leaf scars broadly heart-shaped. Slender shrubs of talus slopes and shaded rimrock in mixed oak-hickory forests. The range for the election pink and the white azalea, *R. oblongifolium,* are similar and shown by the map. *Ericaceae.*

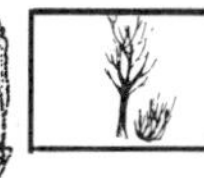

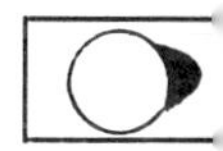

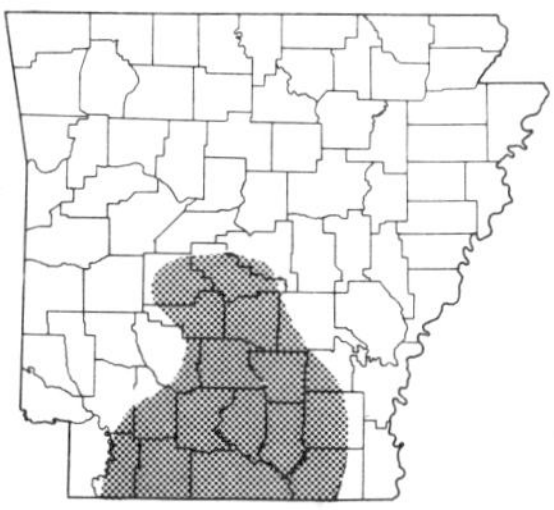

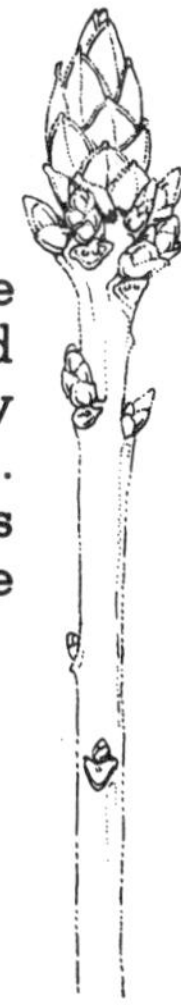

SWAMP AZALEA — *Rhododendron canescens.* A small shrub clumped along margins of lakes, springheads, and streams of coastal sands and gravels. Branch ends often splayed by growth of five to ten twigs. Vegetative buds bunched around a large flower bud at the twig tip. Twigs scurfy. Buds hoary with a gray pubescence. Buds imbricated. Bud scales ciliate, bordered in black, and keel-tipped. Leaf scars alternate, broadly heart-shaped with a single vascular bundle scar. *Ericaceae.*

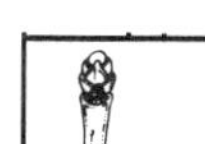

BLACK HUCKLEBERRY — *Gaylussacia baccata.* A single report and collection of it in Montgomery County makes this a rare shrub in Arkansas. Moist woods around spring heads support its growth. Its stiff, dull, reddish-brown branches form short dense shrubs that resemble the blueberries. A terminal bud-scar peg is with each pseudo-terminal bud. Twigs are zig-zagged and clothed in long white hairs sparsely and irregularly distributed. Buds are alternate and loosely imbricated with few scales, the outer scales ribbed and long-tipped. Leaf scars are half-round and fringed with a line of long white hairs. The leaf scar is elevated and rimmed by a ledge. Find a clay-colored or relatively dark pith. *Ericaceae.*

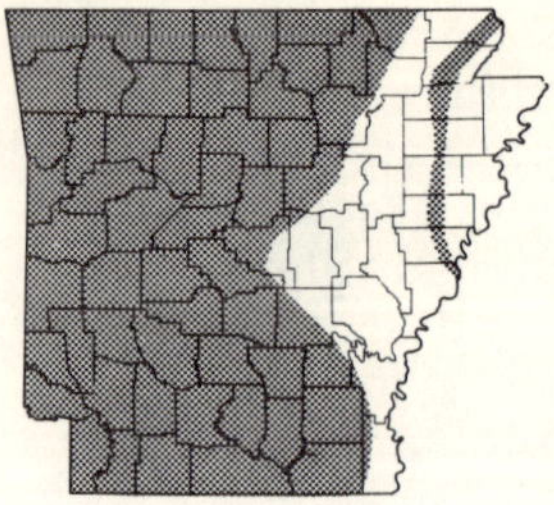

FARKLEBERRY — *Vaccinium arboreum.* A coarse, stiff shrub or small dense-topped tree with tight, thin-shredding bark shows highlights of dark red-brown. Twigs are very slender, hard, reddish-brown, irregular, ridged, and twisted. Protruding leaf scars have minute lateral buds. The black berries, persisting into winter, are hard, dry, stony, and essentially inedible. This shrubby tree is widespread on arid slopes, savannas, and open woods as single trees, clumps, or impenetrable thickets. *Ericaceae.*

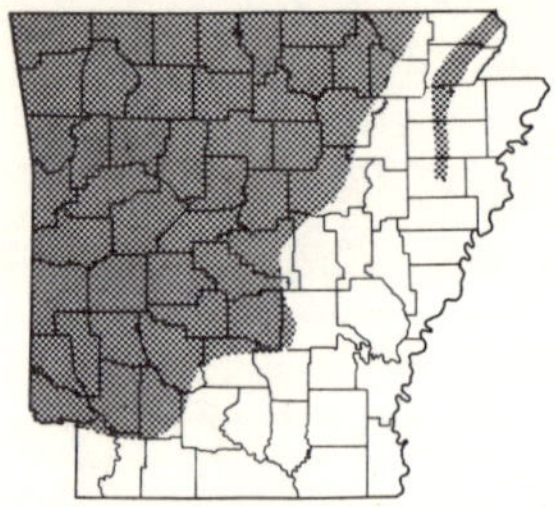

LOWBUSH HUCKLEBERRY — *Vaccinium pallidum.* Flower buds and vegetative buds are imbricate and shiny red. Bud scales have an abrupt, keeled, and acuminate tip. Twigs are green and hairy with a line of denser white hairs running down the stem from the corner of each half-round leaf scar. Chest- to head-high clumps often show vigorous suckers from the base or water-sprouts from arching stems. *Ericaceae.*

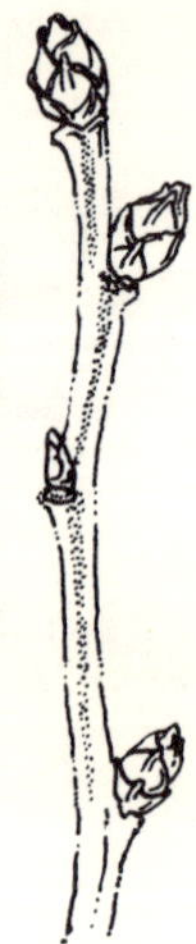

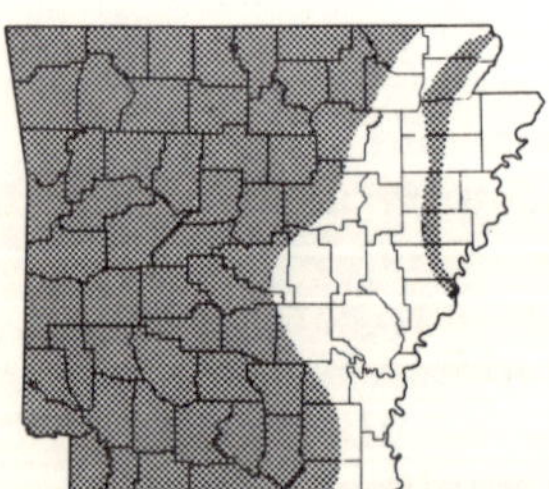

DEERBERRY — *Vaccinium stamenium.* Blueberries and huckleberries are cosmopolitan members of the low canopy communities in Arkansas forests. Their knee-, chest-, and head-high clones command immediate attention. Deerberry is common to well-drained soils. Its supple, diffuse stems grow in knee-high clumps or wide colonies. The pseudo-terminal bud is brown, sharp, and long-pointed. The half-round leaf scars and lateral buds are alternate on slender, glabrous, green twigs. *Ericaceae.*

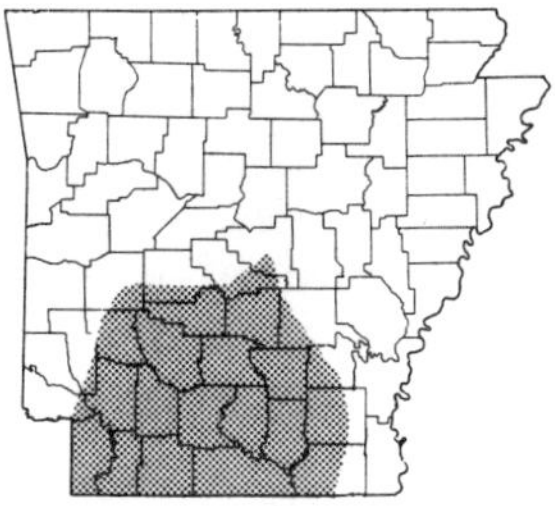

DUSKY BLUEBERRY — *Vaccinium fuscatum.* A head-high bush with a spreading crown conspicuous with its bright purple branches and fat flower buds. Older branches blotched, streaked, and shredding off gray and brown. Suckering is common. Pseudo-terminal buds often are flower buds. Buds are alternate and imbricated with long-pointed and keeled bud scales. Leaf scars are raised crescents fringed with white hairs and have a central patch of vascular bundle scars along their top margin. Found along banks of sandy lakes, stream margins, swamps, and marshes. Mayberry, *V. elliottii,* has a similar range and habitat. *Ericaceae.*

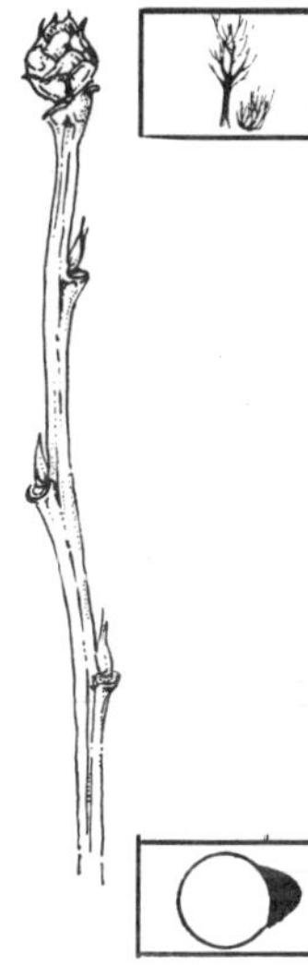

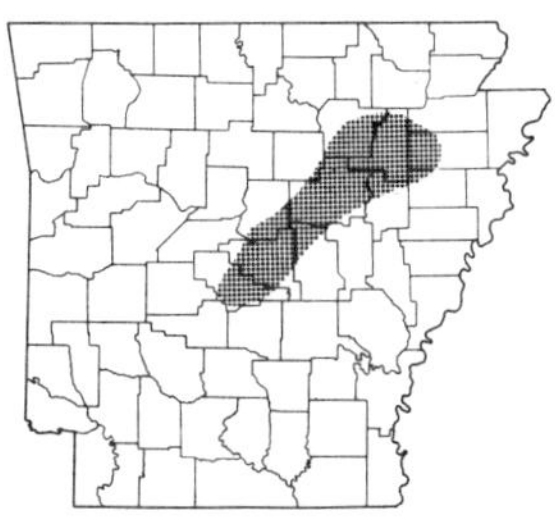

BLACK ALDER — *Ilex verticillata.* A black, multi-trunked shrub, with fine textured ash-gray twigs, of damp thickets and bluff bases. Terminal bud scar is inconspicuous next to the pseudo-terminal bud. Buds are alternate, blunt and imbricated with straw-colored bud scales. Leaf scars are raised, half-elliptical with a central patch of vascular bundle scars. Find dark, withered, stipular tags at the corner of each leaf scar. One year old branches are black with gray superficial flakes. *Aquifoliaceae.*

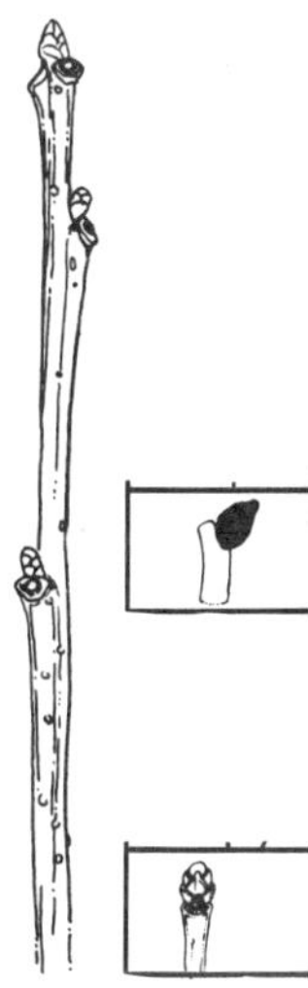

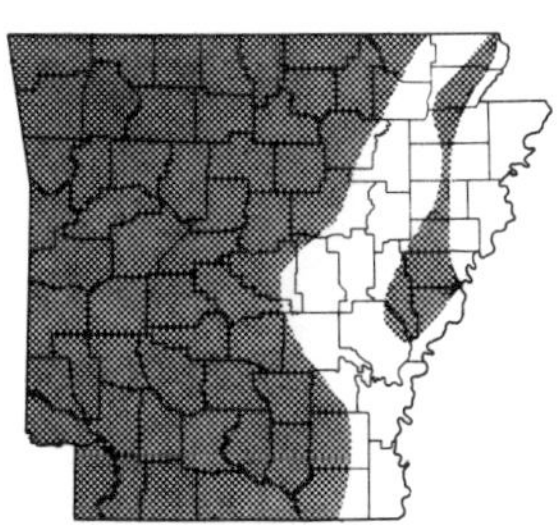

SPICEBUSH — *Lindera benzoin.* Pseudoterminal bud imbricated with a few glossy, bright-green scales. Lateral buds alternate and often with globose, stalked, flower buds. Leaf scars half-elliptical. Twigs bright green, glossy, and aromatic when crushed. A shrub of shaded, moist soils in mixed hardwood forests. *Lauraceae.*

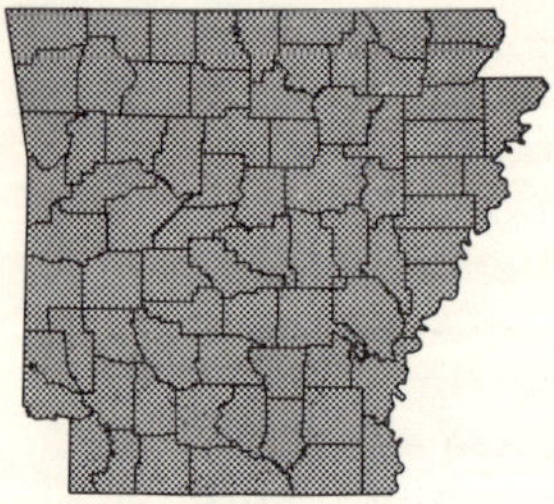

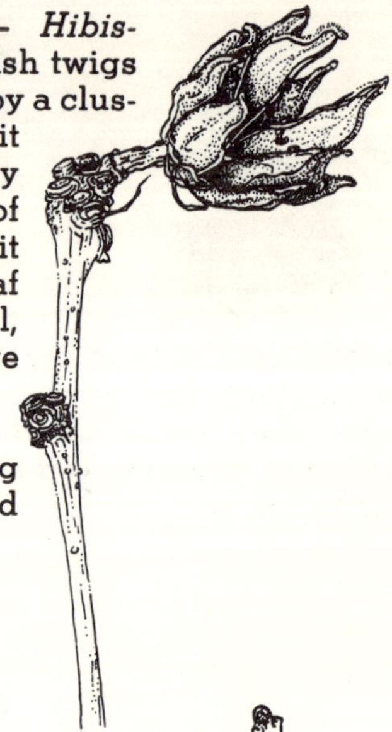

ROSE-OF-SHARON — *Hibiscus syriacus.* Dull reddish twigs are usually terminated by a cluster of flower and/or fruit scars and prominently zig-zagged because of multiple flower or fruit scars at the nodes. Leaf scars are half-elliptical, raised, and attended by stipular scars. A large shrub or small tree with dull streaked bark and many dry, papery, five-celled fruits splitting open at the tips. Seeds have a thin fringe of long blond hairs. Introduced from east Asia and used widely as an ornamental. *Malvaceae.*

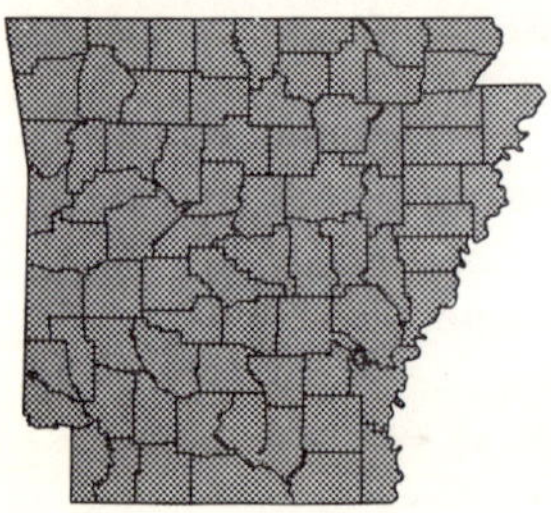

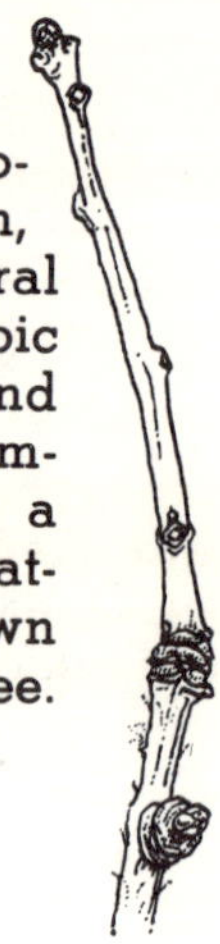

GINKGO — *Ginkgo biloba.* Pseudoterminal bud bright reddish-brown, imbricate, short, and rounded. Lateral buds minute, alternate, above a rhombic leaf scar. Twigs gray to light tan and finely shredded. Dwarf branches common. Excurrent branching produces a straight main trunk with numerous lateral branches. A tree of China grown mostly as an ornamental or street tree. *Ginkgoaceae.*

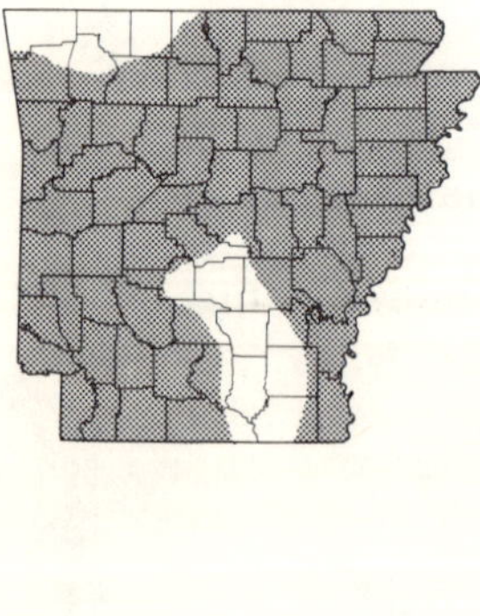

RIVER BIRCH — *Betula nigra.* Twigs wiry with long straight internodes on drooping branches. Buds imbricate, shiny, long-pointed and curved at the tip. Old bark separates away into thin, curled, papery strips showing the salmon-pink underbark. Cone-like fruits can be found on mature trees. Conspicuous trees lining streambanks and water courses. *Betulaceae.*

BLACK GUM — *Nyssa sylvatica.* Twigs slender, smooth, dull, and reddish-brown, supple and easily bent without breaking. Buds imbricate, reddish and downy-tipped. Leaf scars broadly obtriangular with three with three vascular bundle scars. Of upland forest slopes and intermittent streams. *Nyssaceae.*

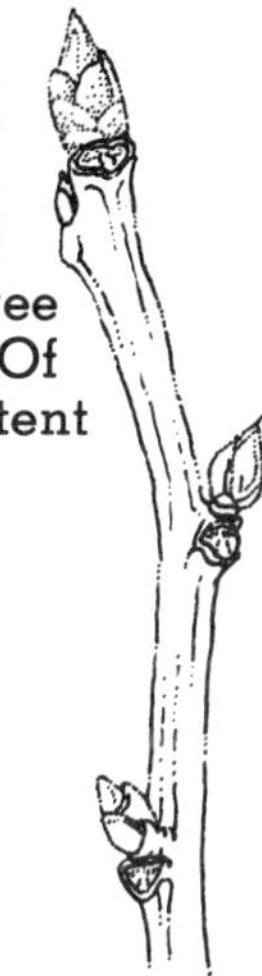

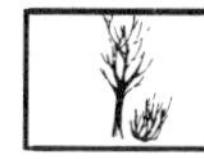

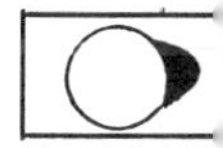

SWEET GUM — *Liquidambar styraciflua.* Pseudo-terminal bud prominent, plump and sharp-pointed, imbricated with glossy, red-brown scales. Leaf scars half-elliptical with three light-colored vascular bundle scars and with faint stipular scars. Twigs glabrous, tan to greenish, often with corky ridges. The stiffly formal excurrent growth produces herringbone patterns in thick populations of young trees. Branches are light ash-gray with golden twigs. Find sharp gumballs under old trees. Pioneer trees in wet oldfields and floodplains. *Hamamelidaceae.*

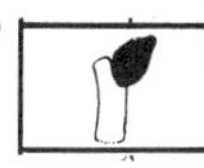

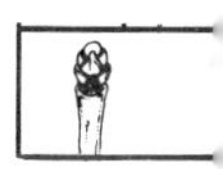

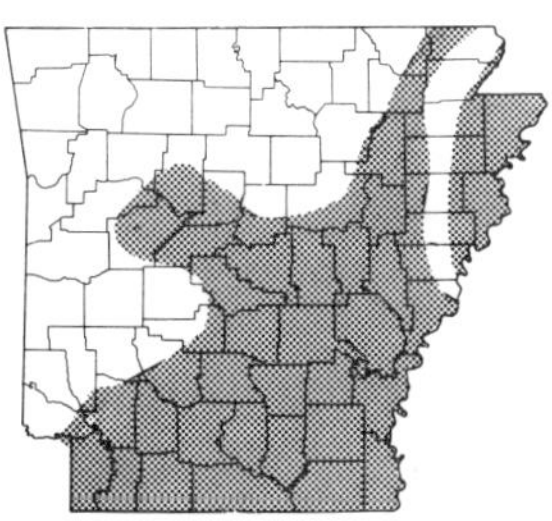

TUPELO GUM — *Nyssa aquatica.* Twigs curved, olive to gold tinted, with leaf scars and terminal bud bunched at the tip. Pseudo-terminal bud dark, imbricate. Alternate lateral buds valvate, smooth, squat, conical, in the notch of a heart-shaped leaf scar. Leaf scar bright with three conspicuous groups of vascular bundle scars. Tall, slender trees with whorls of branches at long intervals. Thumb-sized, bluish fruits resembling olives hang from the limbs or fall, float, and are washed up into drifts. The obviously swollen butts of its trunks are a contrasting form with other types of trees. Found with bald cypress and buttonbush in swamps and along streams. *Nyssaceae.*

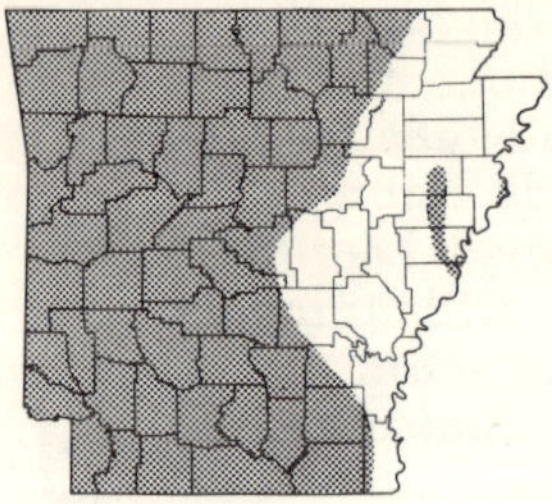

BASSWOOD — *Tilia americana.* Plump, lop-sided, imbricate, shiny, bright orange to red buds on a stout, glabrous stem. Leaf scar half-elliptical with stipular scars. Find several nut-like fruits attached to a slender, leafy wing on or under mature trees. Bark green-gray on young limbs becoming fissured into narrow scaly ridges on the old trunk. Of well-drained moist woods. *Tiliaceae.*

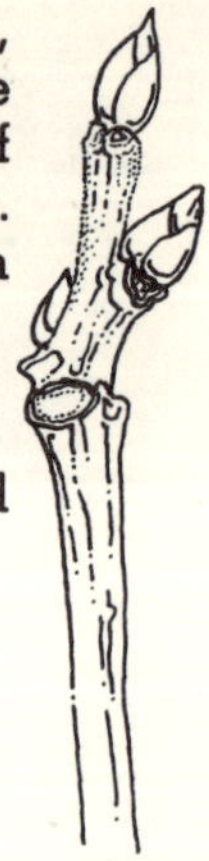

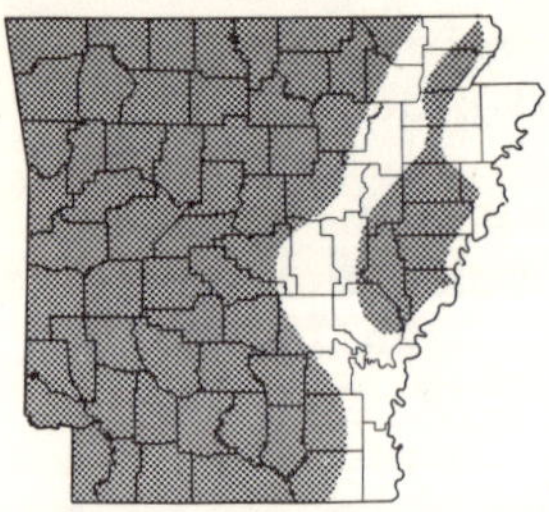

SASSAFRAS — *Sassafras albidum.* Pseudo-terminal bud imbricated with green to yellowish, glabrous bud scales. Lateral buds and leaf scars alternate. Leaf scars raised and half-round. Twigs smooth, shiny, bright green, and aromatic when bruised. The growth form is set by the growth relationship of the pseudo-terminal bud and a single lateral bud down the twig. Find a single vigorous lateral twig whose growth exceeds and finally dominates the original twig. Of old-fields and disturbed open woods. *Lauraceae.*

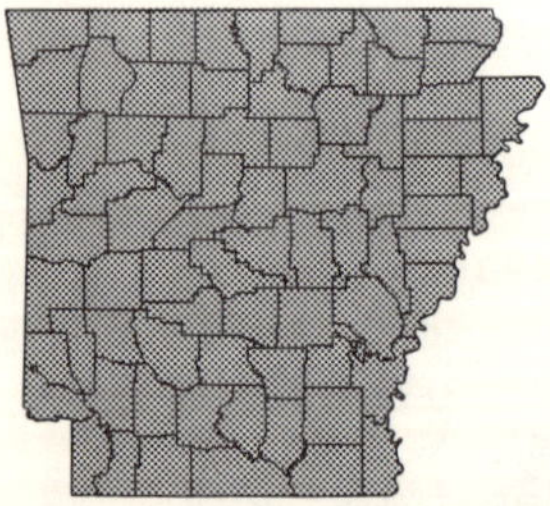

WILD BLACK CHERRY — *Prunus serotina.* Pseudo-terminal bud longer than broad, imbricate, reddish. Lateral buds alternate. Leaf scars raised, half-round with stipular scars. Twigs with pungent odor and astringent taste when crushed. Twigs reddish-brown with prominent lenticels. Lenticels become horizontal streaks on branches and bark. Bark black, flaky with fluted edges. A tree that indicates disturbances in wood's soils. *Rosaceae.*

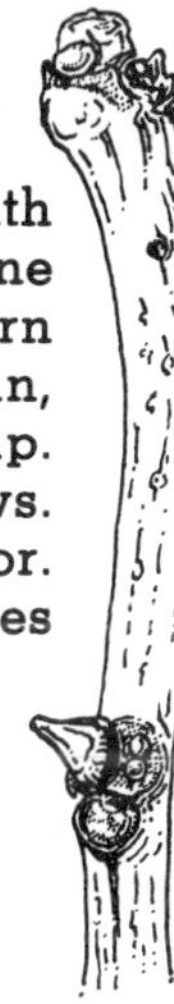

BOIS D'ARC, OSAGE ORANGE — *Malcura pomifera.* Pseudo-terminal bud present. Lateral buds alternate and imbricated. Leaf scars round with collateral and superposed buds. One collateral bud often gives rise to a thorn on vigorous twigs. Twigs gray to tan, smooth and shiny, with milky sap. Branches with natural arches or bows. Bark fissures and exposed roots show a bright orange color. Fruit a large, milky, pulpy, "horse-apple." Dwarf branches present on two-year-old stems. Native of blackland prairies of southwest Arkansas and western Ozark mountains. *Moraceae.*

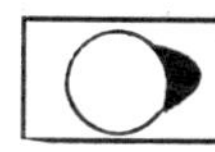

RED MULBERRY — *Morus rubra.* Twigs slender, smooth, brown, and exude milky juice when cut. Buds light brown, imbricate, lustrous, and broad as long. Leaf scars broadly and bluntly obtriangular with many vascular bundle scars. Old bark thin, brown, and tinged with yellow. Root bark yellow. A tree of rich, moist soils in mixed hardwood forests. *Moraceae.*

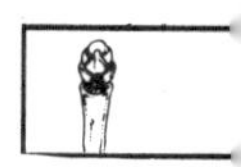

REDBUD — *Cercis canadensis.* Pseudo-terminal bud imbricate and triangular. Lateral buds, obtuse, alternate, often with collateral or superposed, globose flower buds. Upper edge of leaf scar is fringed. Twigs dark, shiny, and zigzagged. A small, understory tree of upland forests. *Leguminosae.*

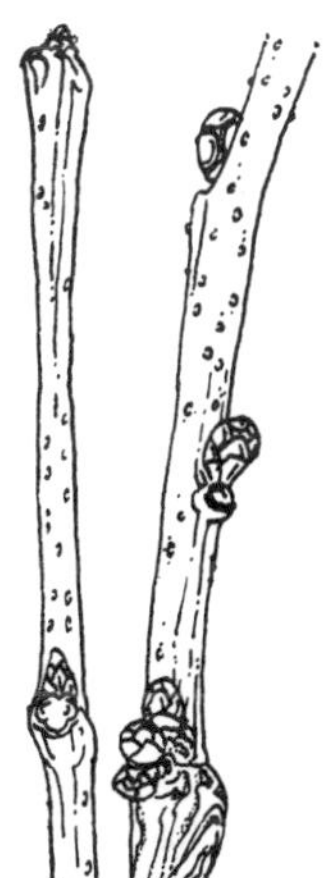

SNOW WREATH — *Neviusia alabamensis.* A fine textured shrub developing a tangled, profuse crown by vigorous, lateral branches. These light brown, lustrous twigs contrast with the old stems, which are purple-gray. The twigs are bowed, twisted, zig-zagged, with dark buds, and wiry toward the tip. Alternate lateral buds are imbricate, squat, sharp, and point outward. Leaf scars are raised, generally rhombic in outline, with three vascular bundle scars, and associated with three ridges that extend down the twig. These ridges become corky, white lines on old stems. Remnants of old fruits are long-stalked with five-parted bases. A rare shrub forming dense, low thickets on talus slopes and dry ridges. *Rosaceae.*

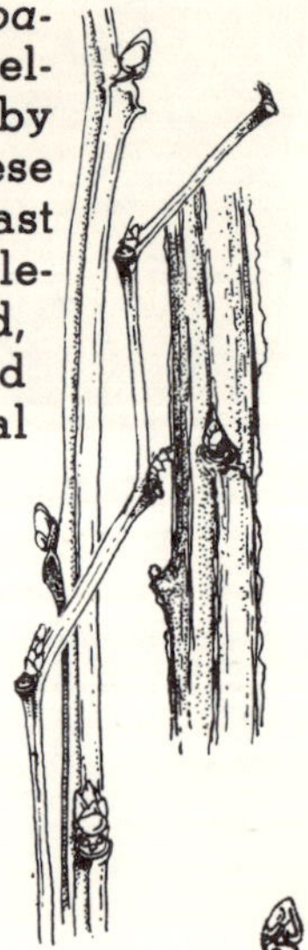

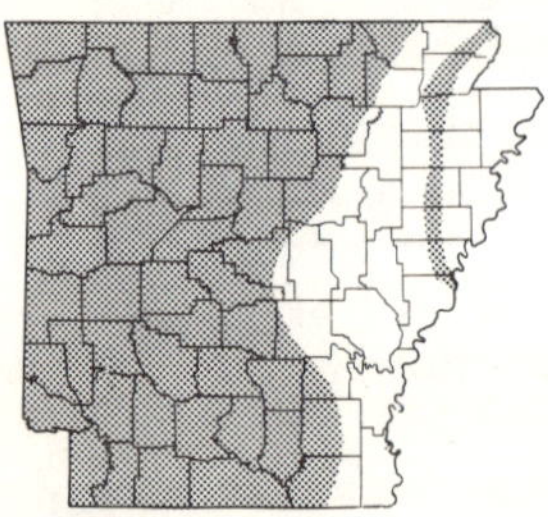

HAZELNUT — *Corylus americana.* Twigs dull brown, zig-zagged with curved internodes. Rounded buds are imbricated and above a relatively small half-round leaf scar. Catkins often present. Fruits, the American Filbert, are enclosed in a papery, leafy, involucre. Shrubs of moist woods, clearings, and thickets. *Betulaceae.*

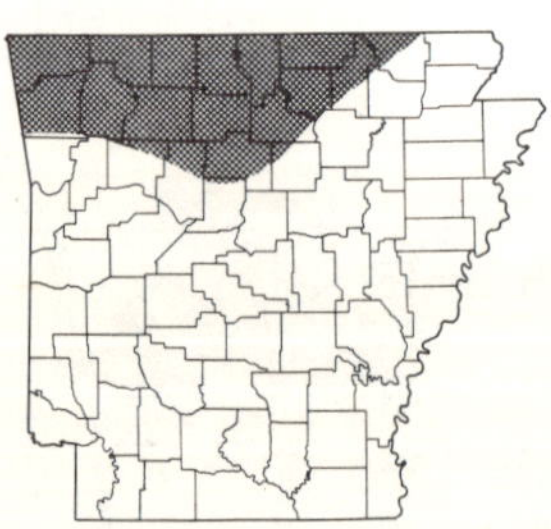

BUCKTHORN — *Rhamnus lanceolata.* Glossy, bright green twigs are zig-zagged with inconspicuous alternate buds. Twig tips are blunted by a terminal leaf scar and superposed buds. Two valvate buds of unequal size are superposed above leaf scars that are shaped like the outline of a kidney bean. Three white vascular bundle scars contrast with the transparent surface of the leaf scar. Faint streaks at the high corner of each leaf scar represent the stipular scars. Older stems show a chambered pith. Diffuse shrubs of moist uplands. *Rhamnaceae.*

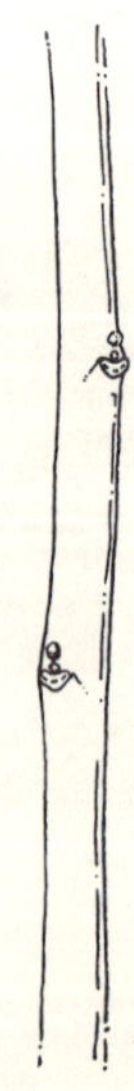

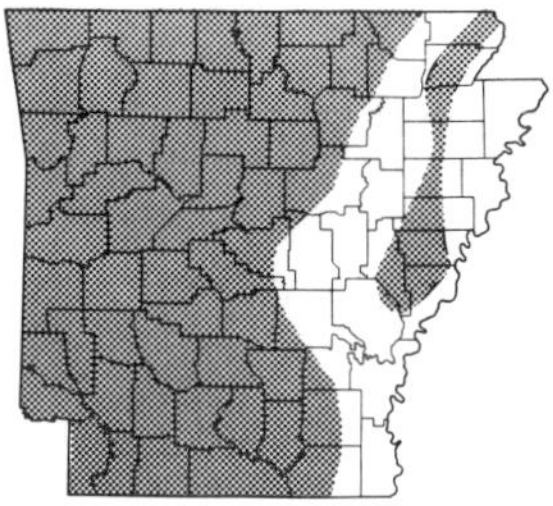

WILD PLUM — *Prunus americanus.* Twigs are glossy, strongly zig-zagged with indented leaf scars and have imbricate buds broadest at base and tapering to a sharp point. Stipular scars are present. A small tree with a spreading crown producing thickets by root-shoots. Cultivated for the red plums and spreading from house sites to roadsides and oldfields. Smith (1978) suggests that *P. americanus* and *P. mexicana,* the big tree plum, intergrade. The map shows a single range for both. *Rosaceae.*

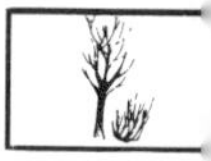

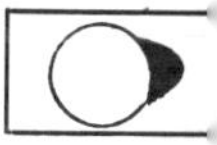

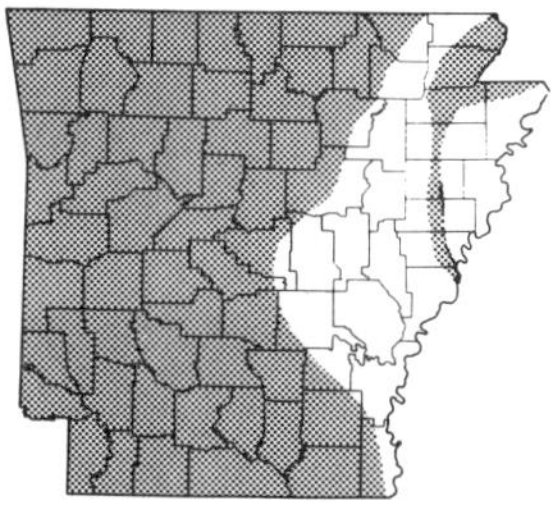

HOP HORNBEAM, IRONWOOD — *Ostrya virginiana.* Slender, zig-zagged twigs have a finely shredded surface and elevated leaf scars with three vascular bundle scars. Bud scales are green, striated and brown-tipped. Old bark is brown, thin, and shredded, strongly resembling the bark of red cedar. A small understory tree of moist soils and floodplains. *Betulaceae.*

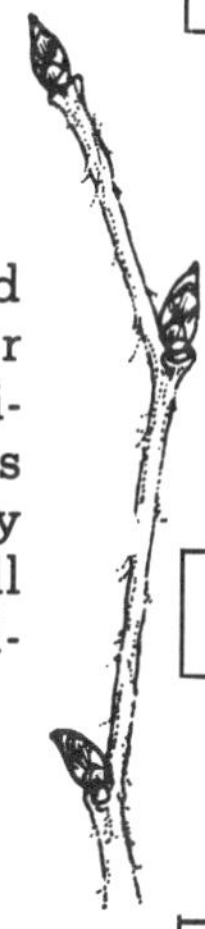

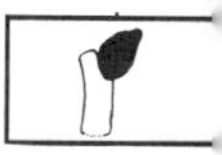

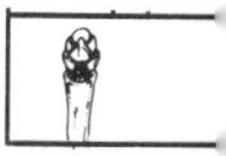

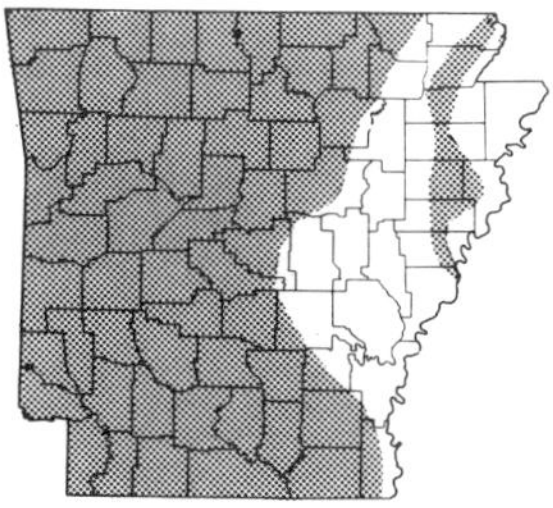

AMERICAN HORNBEAM — *Carpinus caroliniana.* Brown, smooth, zig-zagged, slender twigs. Buds imbricate, brown and downy-tipped. Old bark smooth, bluish-gray, fluted and musclely. Its appearance inspires the common names "blue beech" and "musclewood." A small understory tree of moist soils and low floodplains. Find bunched, leafy fruit clusters. *Betulaceae.*

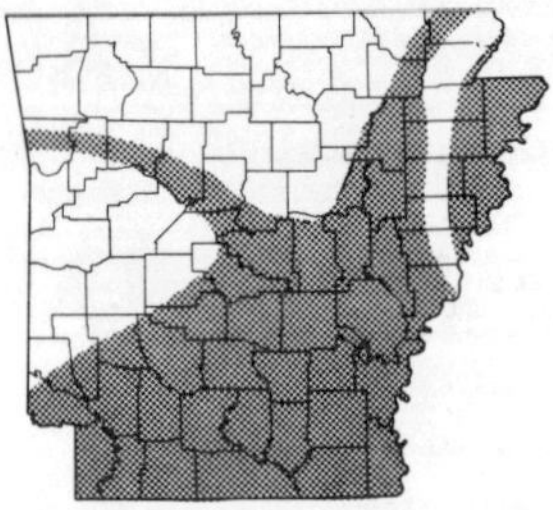

PLANER TREE — *Planera aquatica.* Twigs slender, zig-zagged, scabrous, and dull. Buds alternate, hairy, and superposed. The numerous, spreading branches form a low, broad-crown on this small tree. The short trunk has reddish-brown bark that is scaly and flakes off. A shrubby elm-like tree forming dense stands in swamps and stream habitats. *Ulmaceae.*

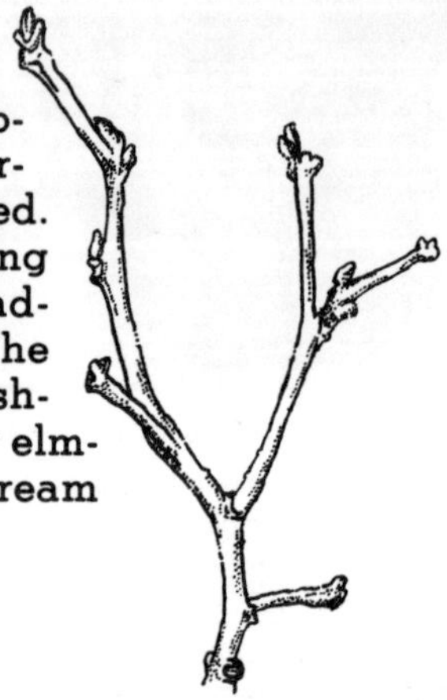

RED ELM, SLIPPERY ELM — *Ulmus rubra.* Pseudo-terminal bud present. Leaf scars and buds are alternate. Buds acute with sharp tips, dark, oblique, and imbricated with reddish, pubescent scales. Twigs reddish-brown, pubescent, dull, and zig-zagged. Branches ascending and diffuse in a broad, fan-shaped crown. Outer bark without alternate light and dark layers, irregularly shedding. Inner bark mucilaginous. Found with mixed hardwoods on moist slopes and high flood-plains. *Ulmaceae.*

AMERICAN ELM — *Ulmus americana.* Pseudoterminal bud present. Buds alternate, oblique, and imbricated with acute, blunt tips. Buds and twigs shiny or only slightly pubescent. Twigs zig-zagged. Branches ascending and diffuse in a broad, fan-shaped crown. Bark gray, ridged and crumbly with alternating layers of light and dark. Found in moist soils with mixed hardwoods. *Ulmaceae.*

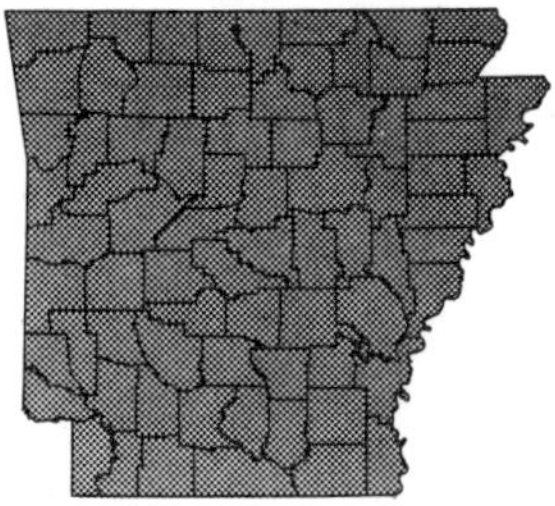

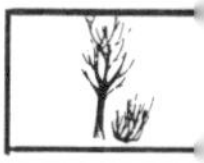

WINGED ELM — *Ulmus alata.* Pseudo-terminal bud present. Buds imbricate, long, pointed, alternate and oblique. Broad, thin, corky wings present on vigorous twigs and branches. The zig-zagged branches are delicate and diffuse. Small trees grow straight and slender. Mature trees show a wide, rounded top. Usually seen as a small tree in dry upland sites. *Ulmaceae.*

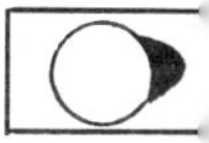

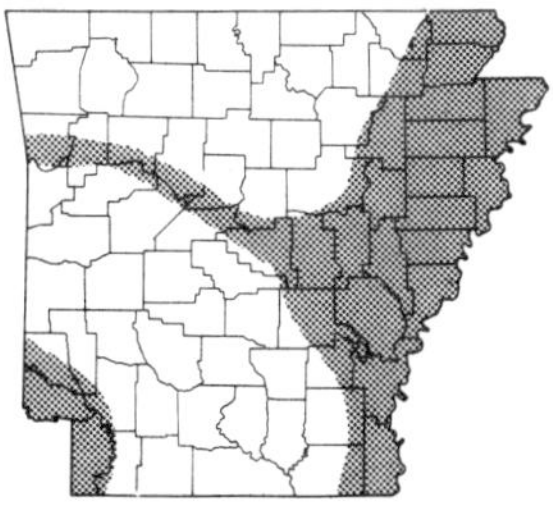

CEDAR ELM — *Ulmus crassifolia.* Twigs and buds pubescent and dull gray. Buds imbricated, sharp-pointed and oblique above lop-sided leaf scars. The delicate, short branches form a dense, narrow crown. Trees of soils frequented by overflow or standing water. *Ulmaceae.*

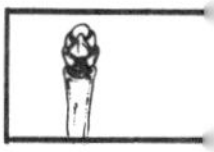

SIBERIAN ELM — *Ulmus pumila.* An introduced, cultivated tree common to the Ozarks and Crowley's Ridge, particularly around farmsteads. Often repeatedly cut back to achieve a coppiced effect. Twigs and branches delicate, zig-zagged, supple, hirsute, and ash gray with dark purple-brown, oblique, alternate buds. Buds round-tipped and imbricated with ciliate scales. Leaf scars half-round to ovate with three dark patches of vascular bundle scars. Pseudo-terminal bud attended by a terminal bud scar. *Ulmaceae.*

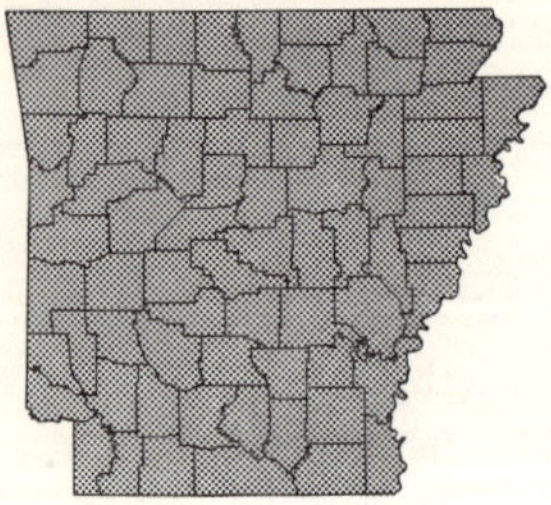

SUGARBERRY — *Celtis laevigata.* Slender, zig-zagged, wiry twigs with tip die-back. Lateral buds imbricate, triangular, and closely appressed to the twig. The half-round leaf scar is elevated. Growth of branches is strongly excurrent with few laterals. Old bark mostly smooth gray with noticeable layered warts. Streamsides and bottomlands. *Ulmaceae.*

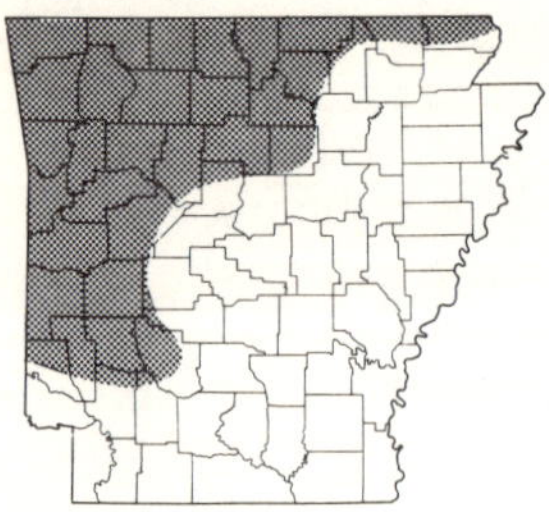

HACKBERRY — *Celtis occidentalis.* Twigs zig-zagged with closely appressed, conical buds, chevron-shaped leaf scars and stipular scars. Witches-brooms fungus common in old trees. A tree of upland slopes and sweet soils. *Ulmaceae.*

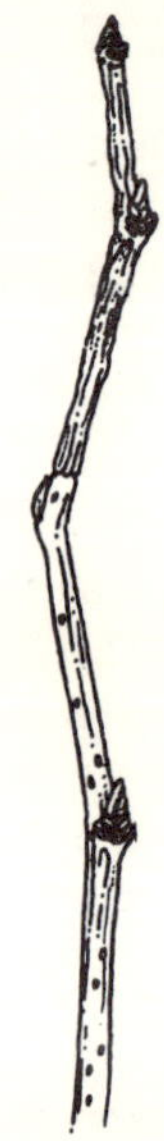

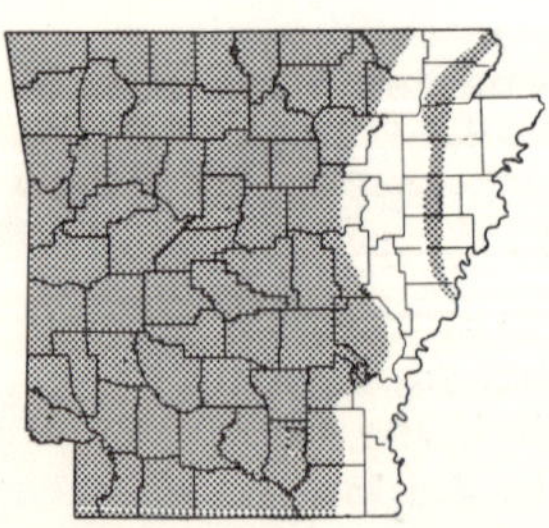

NEW JERSEY TEA — *Ceanothus americanus.* Twigs and buds are dull and rough, brown to gray. Disk-like fruit bases persist into winter, dies back, but is renewed by annual growth. A small delicate erect, shrubby perennial of dry, open woods. *Rhamnaceae.*

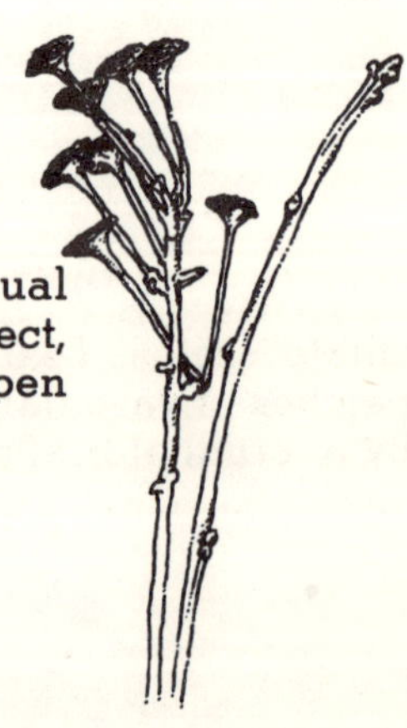

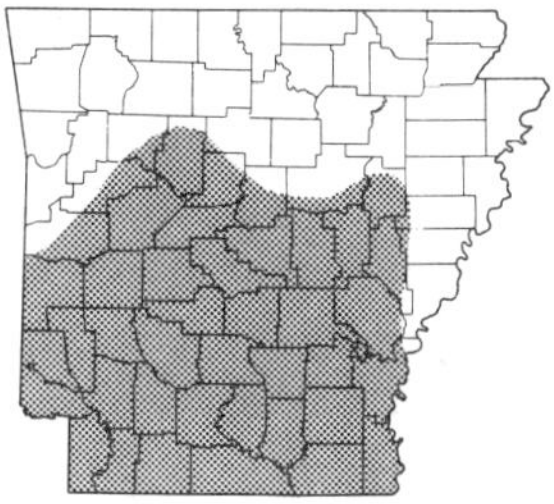

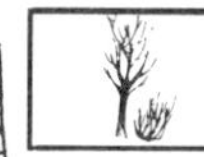

SEA-MYRTLE — *Baccharis halimifolia.* A head-high shrub with gray-brown main stems and delicate, diffuse, light-green twigs. All twigs are terminated by die-back or tiny, daisy-like dried flowers. Twigs are supple, six- to eight-angled and covered with dry, aromatic resin. Alternate buds and leafscars are inconspicuous and located mostly at the base of branches. A few narrow leaves often persist into winter. Abundant in ditches and old fields where gumbo and wet soils prevail. *Compositae.*

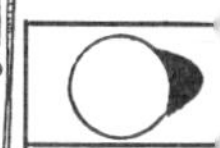

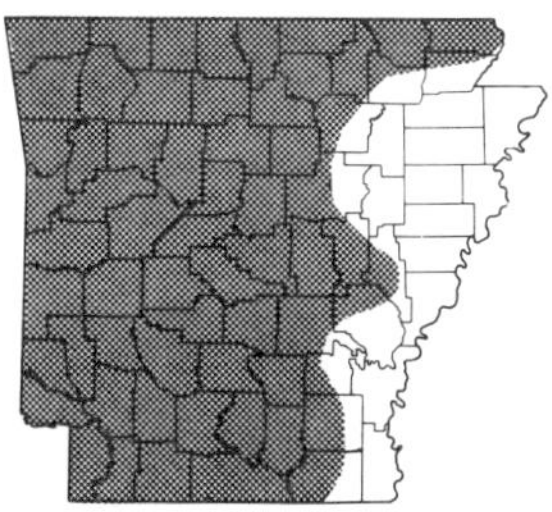

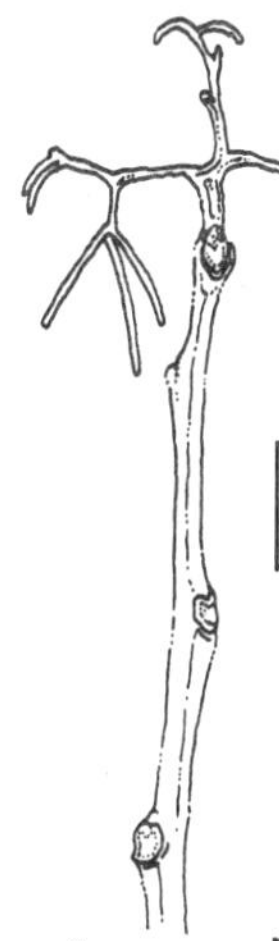

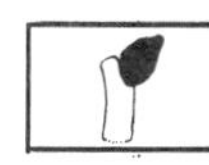

WAFER ASH — *Ptelea trifoliata.* A small, smooth shrub with pungent bark. Remnants of terminal fruit clusters persist into winter. Leaf scar heart-shaped around a silvery, silky bud. Of moist, shaded soils of upland forest. *Rutaceae.*

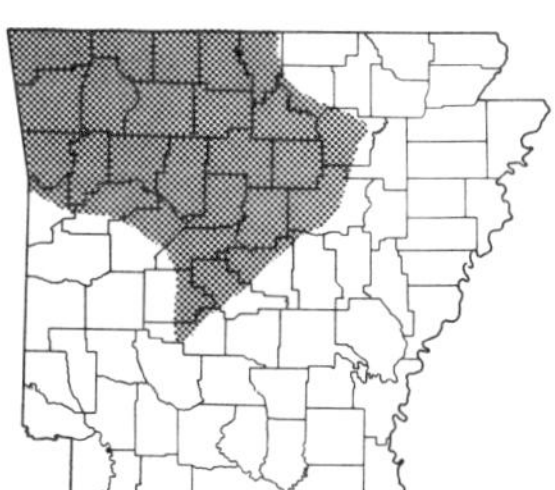

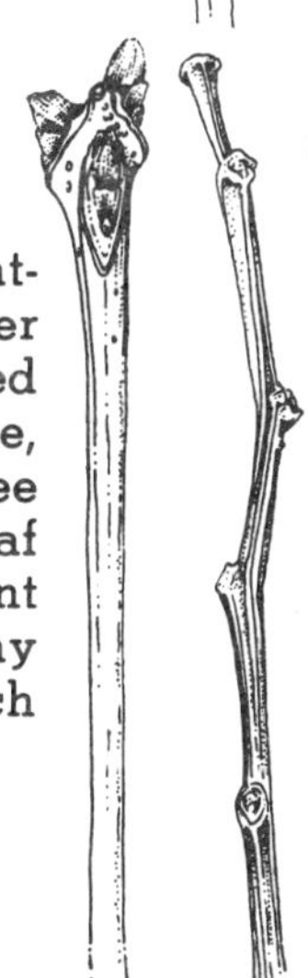

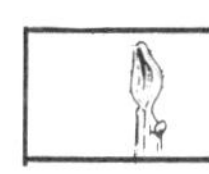

YELLOW WOOD — *Cladrastis kentukea.* Twigs glossy, coarsely zig-zagged and blunt at the tip by a terminal bud scar and a broadly obtriangular leaf scar. Alternate lateral buds are embedded in the center of the leaf scar or stick out as a naked bud covered with dense, coarse, smooth, tawny hairs. Two or three superposed buds of unequal size often emerge from the leaf scar. Wood is canary yellow when fresh cut and has the faint taste of green peas. Older branches have a thin, smooth, gray bark. Find thin, papery fruit pods during the fall. Of rich wooded ravines and limestone bluffs. *Leguminosae.*

THORNS

- **Gum Bumelia** : Terminal thorn short. p. 56
- **Bois D'Arc** : Lateral thorn next to accessory bud. p. 57
- **Cockspur Thorn** : Lateral thorn. p. 57
- **Honey Locust** : Lateral thorn, branched. p. 57
- **Water Locust** : Lateral thorn, branched. p. 58

SPINES

- **Black Locust** : Stipular spines curved. p. 58
- **Bristly Locust** : Stipular spines straight. Bristles stiff, brown. p. 58
- **Prickly Ash** : Stipular spines broad-based and spreading. p. 59
- **Missouri Gooseberry** : Stipular spine slender, stout, red. p. 59
- **Prickly Gooseberry** : Stipular spines in multiples, brown. p. 59
- **Carolina Rose** : Stipular spines, stiff. Prickles scattered, flexible. p.60

PRICKLES

- **Red Raspberry** : Prickles thin, straight. p. 60
- **Black Raspberry** : Prickles thin, sparse. p. 60
- **Southern Dewberry** : Prickles and bristles maroon. p. 61
- **High-Bush Blackberry** : Prickles random, thin, recurved. p. 61
- **Multiflora Rose** : Prickles near lateral buds. p. 61
- **Hercules' Club** : Prickles concentrated at leaf scar margin and also occur at random on stout twigs. p. 62
- **Toothache Tree** : Prickles purple, flat, on purple-tipped twigs. p. 62

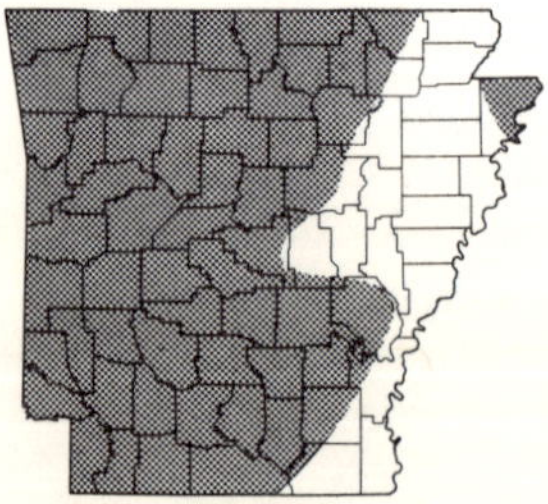

GUM BUMELIA — *Bumelia lanuginosa.* Twigs terminated by a single short, hard thorn, derived from the withered twig tip. Twigs glabrous. Buds downy and pale. Leaf scars half-round and recessed. Trees of exposed rimrock and dry sites. *Sapotaceae.*

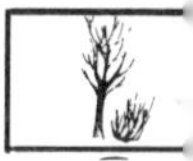

BOIS D'ARC, OSAGE ORANGE

— *Maclura pomifera.* Buds alternate. Twigs with or without thorns. Straight, hard thorns arise from lateral buds, accompanied by a leaf scar and accessory buds. Twigs gray to tan, smooth and shiny, with milky sap. Branches with natural arches or bows. Bark fissures and exposed roots show a bright orange color. Fruit a large, milky, pulpy, "horse-apple." Dwarf branches present on two-year-old stems. Native of blackland prairies of southwest Arkansas and western Ozark mountains. *Moraceae.*

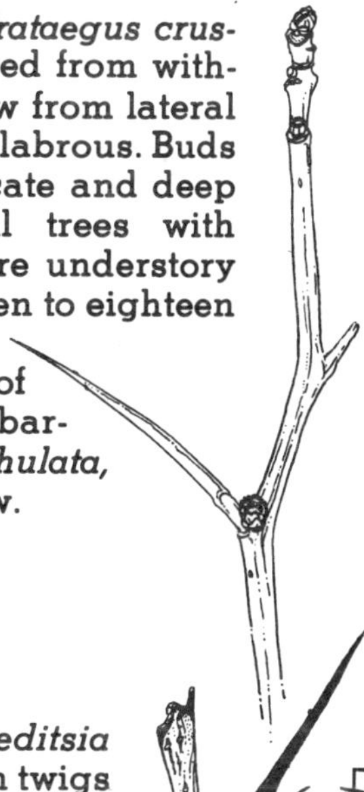

COCKSPUR THORN

— *Crataegus crus-galli.* The thorns are derived from withered lateral twigs that grow from lateral buds. Twigs and buds are glabrous. Buds are blunt, rounded, imbricate and deep red. These thorny, small trees with rounded, broad, crowns are understory trees of moist woods. Sixteen to eighteen species of Crataegus are recognized in the state as members of a wide variety of habitats. The common species are: *C. berberifolia,* barberry leaved haw; *C. marshallii,* parsley haw; *C. spathulata, spatulate-leaved haw; and C. viridis,* the green haw. *Rosaceae.*

HONEY LOCUST

— *Gleditsia triacanthos.* Green to brown twigs with pseudo-terminal buds show conspicuously large, branched thorns. Leaf scars are half-round with three vascular bundle scars. Superposed buds appear above minute lateral buds. The bark is dark and fluted. Long, flat, twisted pods can be found that contain a dozen or more hard beans. A tree often associated with soils abandoned after disturbances by floods, pasturing, or row cropping. *Leguminosae.*

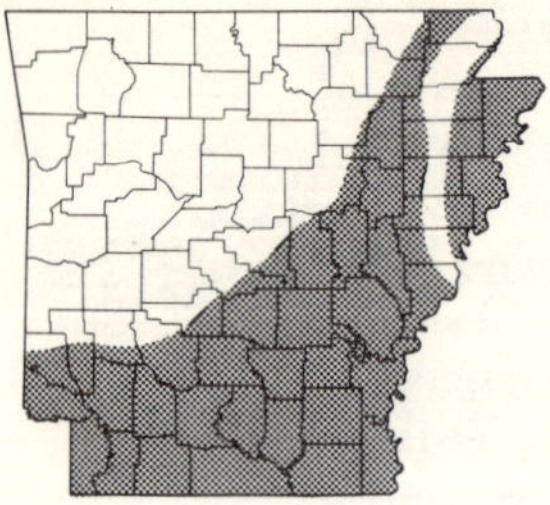

WATER LOCUST — *Gleditsia aquatica.* Green speckled stems with bright-red, flattened, and cross-branched thorns on a tree growing in a swamp or wet woods will identify it. Twigs are blunted by a broad terminal bud scar and a pseudo-terminal bud. Superposed buds are alternate at swollen nodes and line out above a V-shaped leaf scar. Each leaf scar shows three groups of vascular bundle scars. Find straight flat fruit pods, about equal in length to the span of a hand, with one to three seeds without pulp. *Leguminosae.*

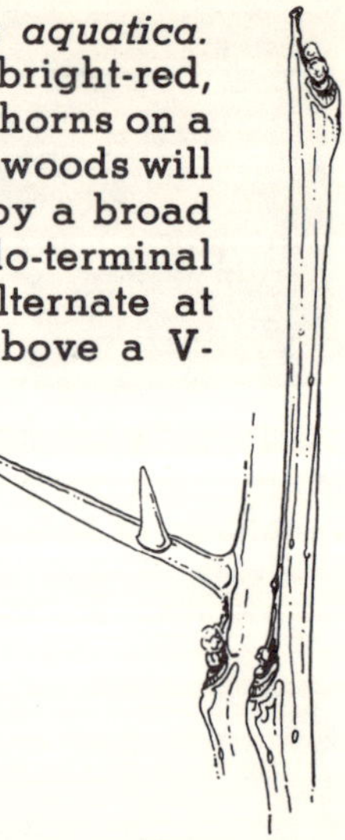

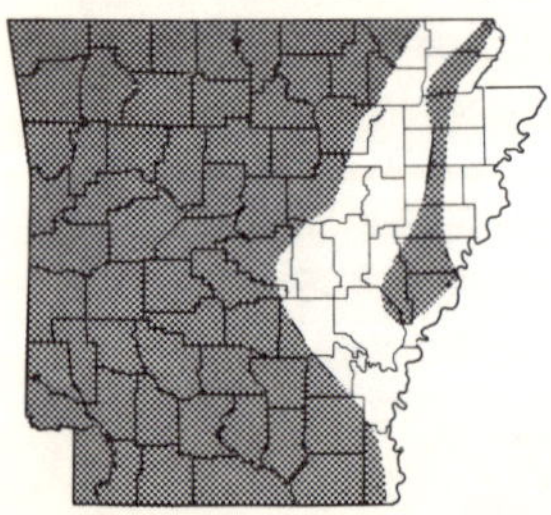

BLACK LOCUST — *Robinia pseudo-acacia.* Buds alternate and embedded in a widely triangular leaf scar. Strong, curved stipular spines occur on either side of the leaf scar. Decurrent lines run down the stem from each angle of the leaf scar. Twigs are zig-zagged, stout, dark, and shiny. Old fruit pods contain four to eight hard beans. Native to the interior highlands, cultivated elsewhere. *Leguminosae.*

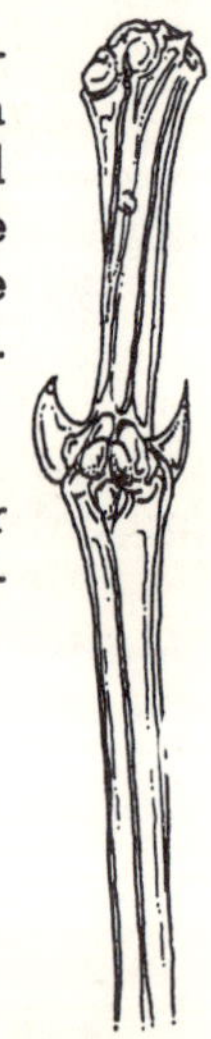

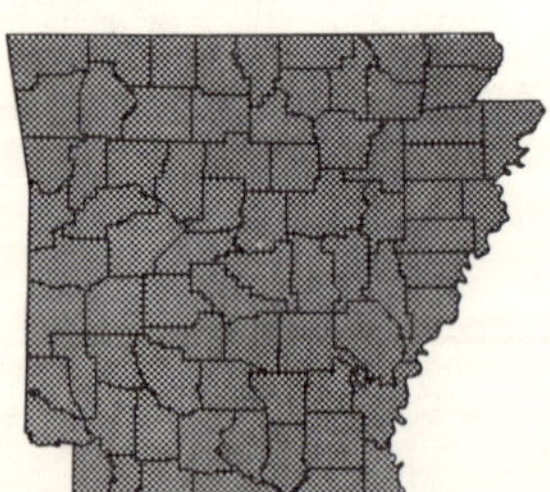

BRISTLY LOCUST — *Robinia hispida* Twigs dull and covered with stiff brown bristles. Straight, weak stipular spines are at the side of each leaf scar. Lateral buds embedded in the leaf scar. Cultivated for its rose-purple flowers. Escaping. *Leguminosae.*

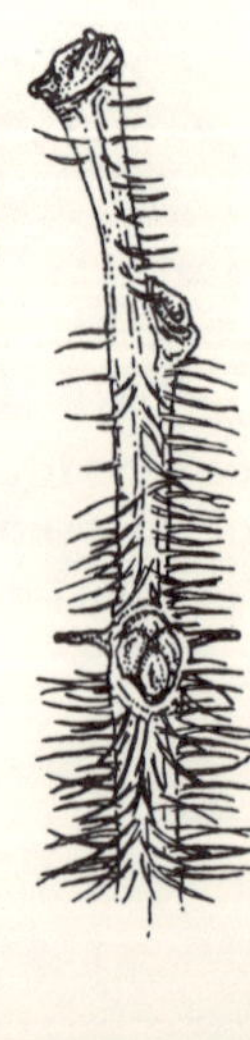

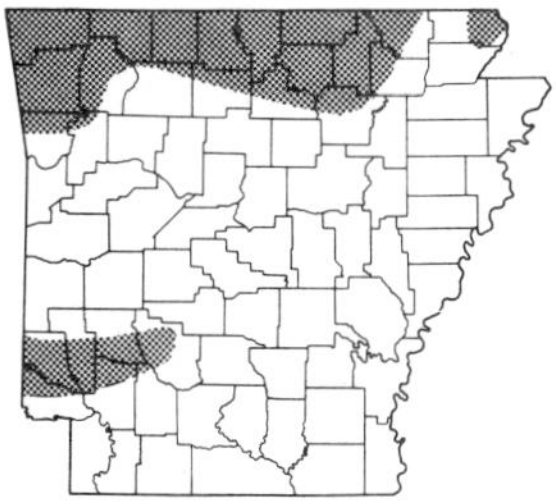

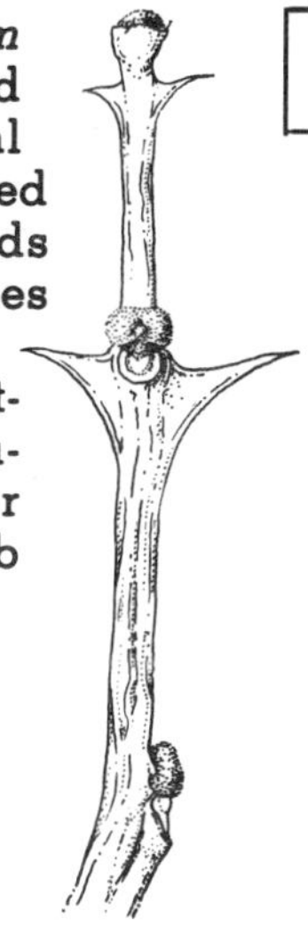

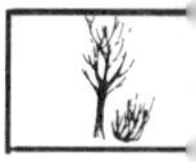

PRICKLY ASH — *Zanthoxylum americanum.* Stipular spines spread at the sides of some leaf scars. Lateral buds alternate, blunt, and covered with dense, rusty-brown hairs. Buds later develop into dwarf branches showing clusters of numerous leaf scars. Leaf scars broadly heart-shaped with three patches of vascular bundle scars. Twigs nearly black with red tones. Older stems dark gray with fine, light streaks. A small tree or shrub in woods of limestone soils. *Rutaceae.*

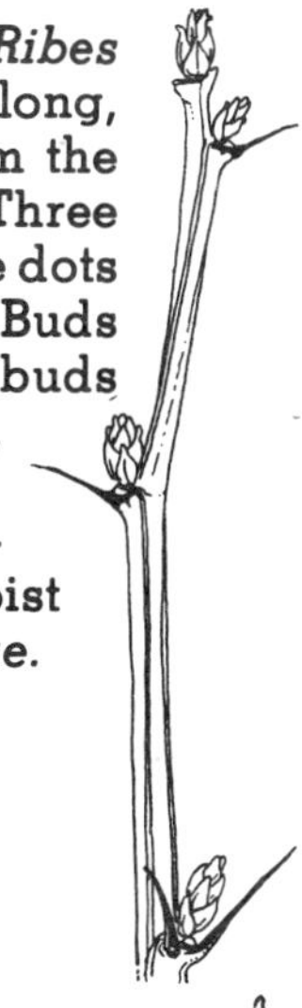

MISSOURI GOOSEBERRY — *Ribes missouriense.* One, two or three long, stout, red stipular spines arise from the margins of the crescent leaf scars. Three vascular bundle scars show as white dots along the top margin of the leaf scar. Buds are alternate. Terminal and lateral buds are loosely imbricated with brown, papery scales. Twigs pale, ridged, and angled. Old stems glossy, maroon, and flaky. Low, fine-textured shrubs of open woods, thickets, and fencerows. Moist upland hardwoods of the Springfield Plateau. *Saxifragaceae.*

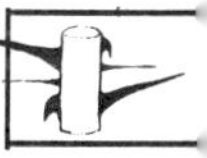

PRICKLY GOOSEBERRY — *Ribes cynosbati.* Multiple spines, brown and round in cross section, arise at the base of the slender, crescent leaf scars. A shrub of rocky slopes and dry woods on Mount Magazine and Rich Mountain. *Saxifragaceae.*

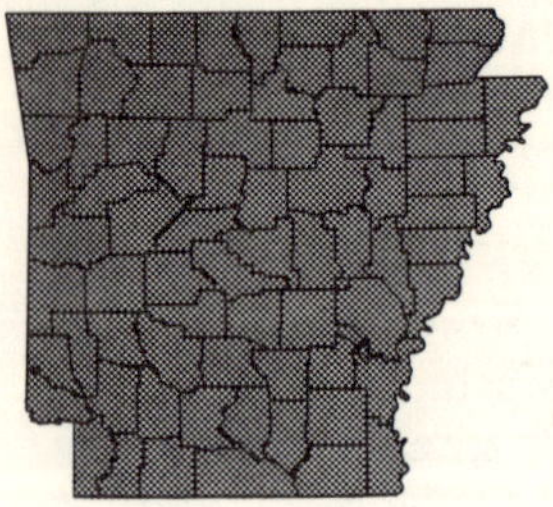

CAROLINA ROSE — *Rosa carolina.* A low, delicate rose with single, slender canes of bright green and maroon. Armed with longer, stiff, stipular spines at each leaf scar and with shorter, flexible, numerous, scattered prickles on the twig internodes. Alternate lateral buds are loosely imbricated with a few bright red scales. Leaf scar is a thin line of a half circle with three vascular bundle scars. Canes end with a terminal bud or a stalked hip. Inconspicuous roses of dry, open habitats in thin woods. *Rosaceae.*

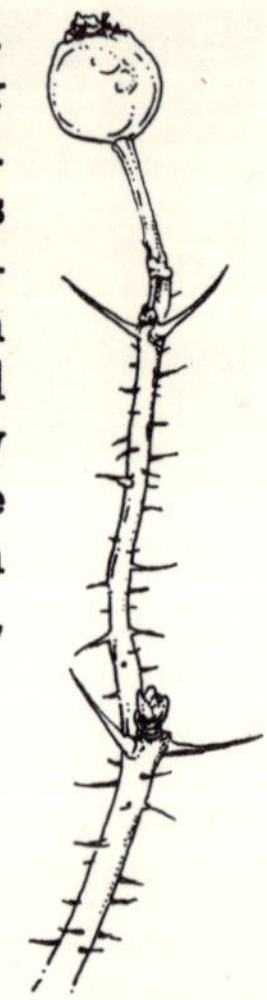

RED RASPBERRY — *Rubus idaeus.* New canes have thin, straight prickles on twigs and old leaf petioles. Stems highly colored, glaucous, and pithy. Die-back common or twigs have a loosely imbricate terminal bud. Alternate buds imbricate and pointed, sitting on a shelf formed by the persistent petiole base. Cultivated and spreading. *Rosaceae.*

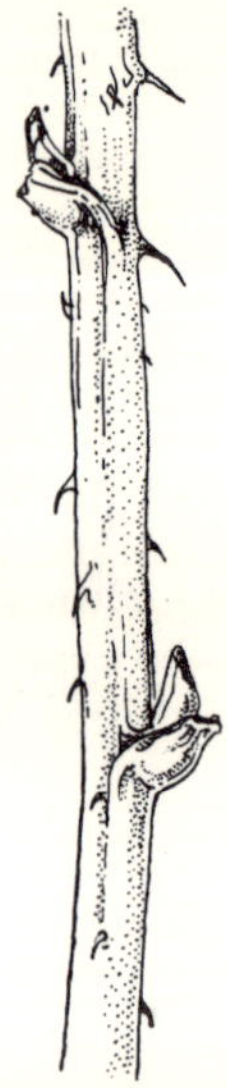

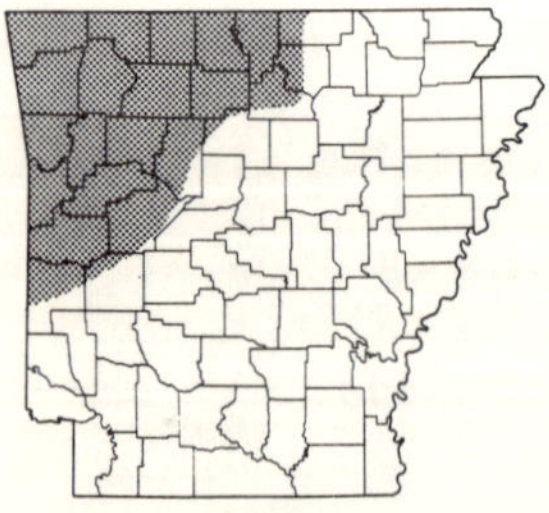

BLACK RASPBERRY — *Rubus occidentalis.* New canes deep maroon, glaucous, and pithy. Prickles thin and sparse on twigs and old leaf petioles. Die-back on canes common. Alternate buds constricted at the base, slightly stalked, and imbricated with many loose, pointed bud scales. Leaf petioles persist. Found in dense colonies that spread along roadsides, thickets, and old house sites. *Rosaceae.*

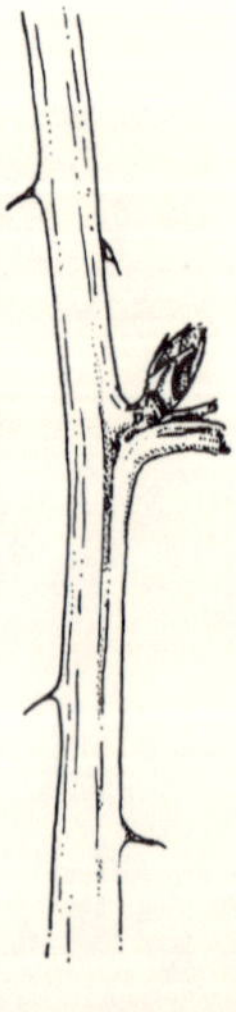

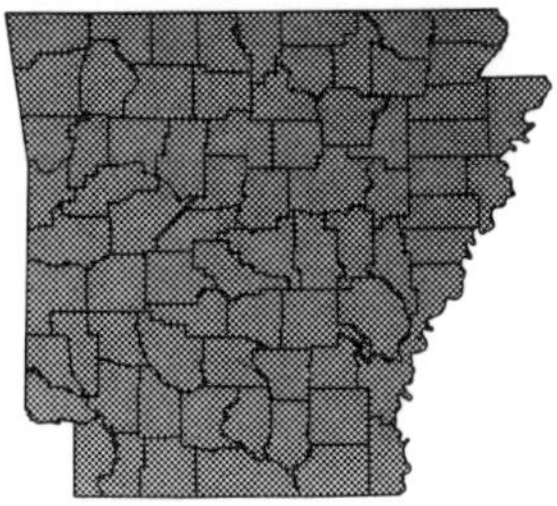

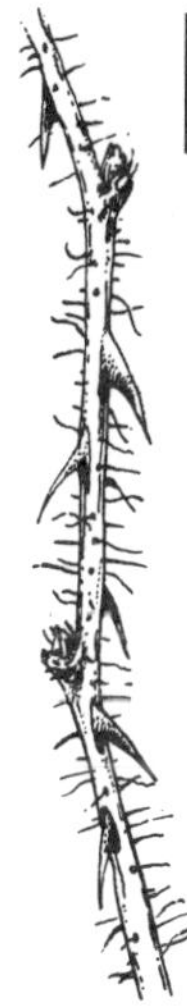

SOUTHERN DEWBERRY — *Rubus trivialis.* Low, trailing, thin, limber canes, retain many leaves into late winter which changed color with the stems. Twigs are covered with maroon prickles and stiff bristles. Prickles recurve at a 45-degree angles from the stems. Ankle- to knee-high patches of maroon brambles are in exposed areas of ditches, beaches, roads, and fields. *Rosaceae.*

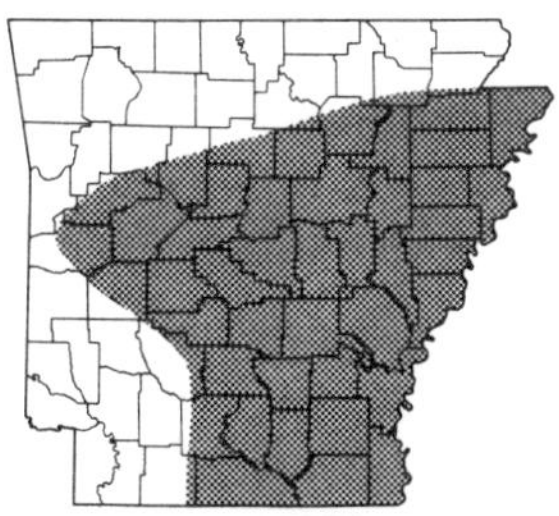

HIGH-BUSH BLACKBERRY — *Rubus argutus.* Long, erect, arching, maroon and green canes are unbranched for a long distance at the ends and show die-back, a pseudo-terminal bud, or late-season leaves. Alternate lateral buds are imbricate, pink with fuzzy, white tips. Leaf scars are covered with a remnant of the leaf petiole base. Twigs zig-zagged with bowed internodes. Old stem ridged; starshaped in cross section. All stems have random, thin, recurved prickles. Seen in winter as chest-high clumps of maroon brambles among grasses of roadsides, fencerows, pastures, open woods, and old fields. *Rosaceae.*

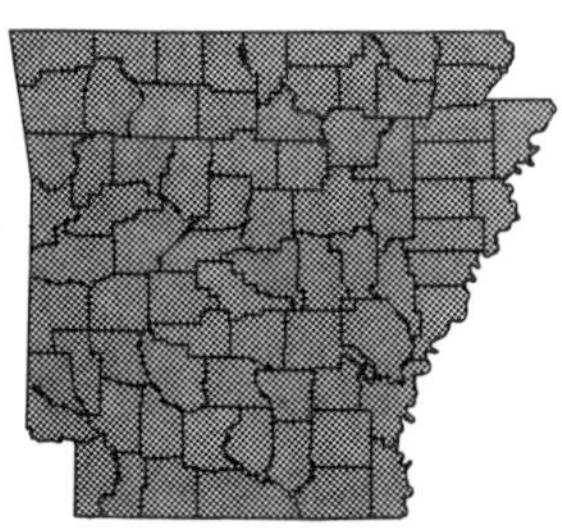

MULTIFLORA ROSE — *Rosa multiflora.* Twigs and branches strongly zig-zagged, green and purple, terminated by a loosely imbricated bud. Alternate lateral buds subtended by two lateral, recurved prickles. Some random prickles occur between nodes. The leaf scar and stipule scar are continuous as a thin line between the bud and prickles for more than half the circumference of the twig. Produces clusters of red, pea-sized hips. Cultivated; escapes to fencerows, old fields, and open woods. *Rosaceae.*

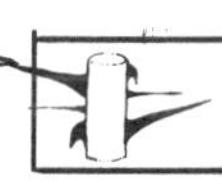

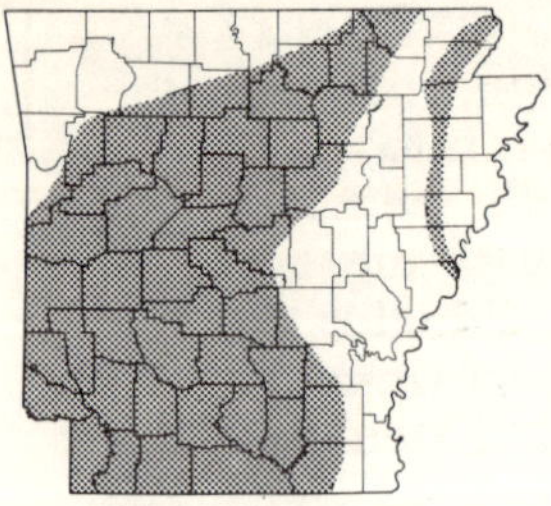

HERCULES' CLUB — *Aralia spinosa.* A prickly, unbranched, small tree with stout brown stems and exceptionally large, crescent leaf scars, which reach nearly around the stem. This tree forms thickets in the low canopy of woods where there are well-drained soils. *Araliaceae.*

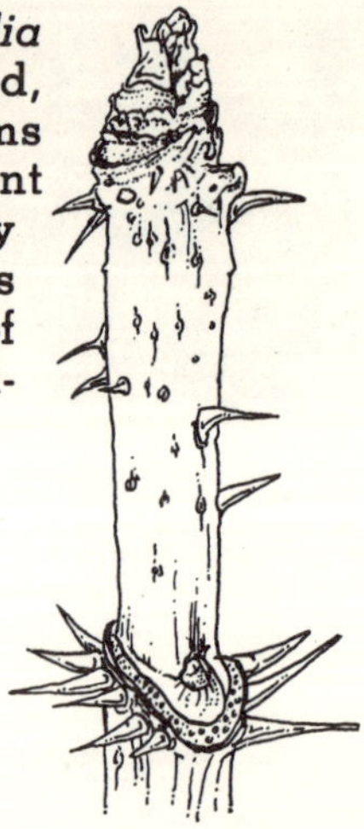

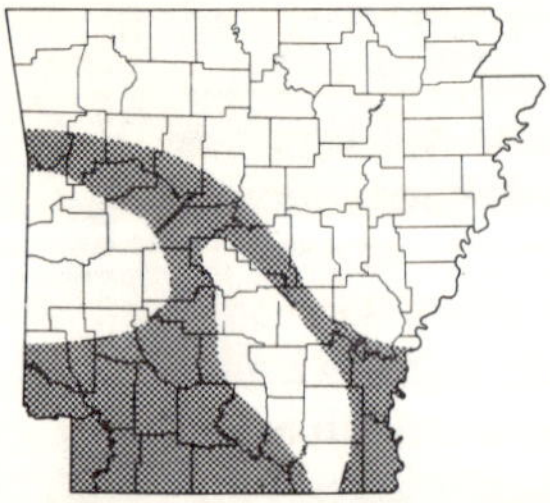

TOOTHACHE TREE — *Zanthoxylum clava-herculis.* Terminal bud withered; low, blunt, purple, and granular. Lateral buds are alternate with raised leaf scars. Leaf scars are either heart-shaped or crescent-shaped with three lobes showing three groups of vascular bundle scars. Twigs have dark purple tips and purple, flat prickles. A wide tree of medium height with wide forks and bowed limbs. Ash-gray old bark is smooth with numerous laminated, wedge-shaped warts, terminated by a long, sharp prickle. Fresh-cut inner bark has numbing effect on lips and gums. Of alluvial soils. *Rutaceae.*

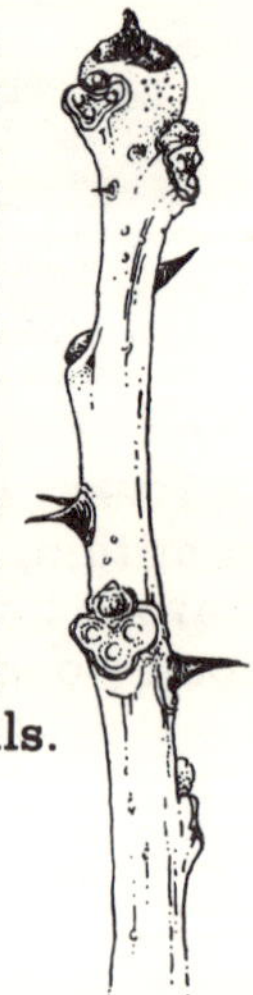

Opposite Buds

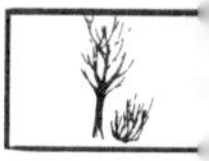

CLIMBING BY AERIAL ROOTS :

Trumpet Creeper : Leaf scar rimmed. Stipular scars connected. Stems brown, downy. Fruits large, 2-parted. p. 63

CLIMBING BY TWINING LEAF PARTS :

Virgin's Bower : Stems delicate, slightly woody. Leaf petioles persist. Fruits fuzz-buttons. p. 63

Crossvine : Petiole crossarm with two leaflet scars and a tendril. Stem 4-angled. Fruits large, flat, dray, parchment-like. p. 64

CLIMBING BY TWINING STEMS :

Climbing Dogbane : Fruits long slender. Seeds beaked. Twigs smooth, brown. Buds covered with a flap. p. 64

Japanese Honeysuckle : Leaf petioles persist. Stem slender, hairy. p.64

Grape Honeysuckle : Buds with chaffy scales. Stipular scars. Stems brittle, smooth, hollow. p. 65

Coral Honeysuckle : Buds with chaffy scales. Stipular scars. Stems brittle, smooth, straw-colored. p. 65

TRUMPET CREEPER — *Campsis radicans.* Light-brown vines, seldom branching, climb by aerial roots at nodes. Twigs are smooth with short, erect, white hairs. Tip die-back present. Lateral buds and leaf scars opposite. Leaf scar rimmed and depressed with a circular pattern of vascular bundle scars. Faint stipular scars are connected. Old stems twisted, with bark shredding into two different colors. One can find hard, elongated, drooping, two-parted fruits with winged seeds. *Bignoniaceae.*

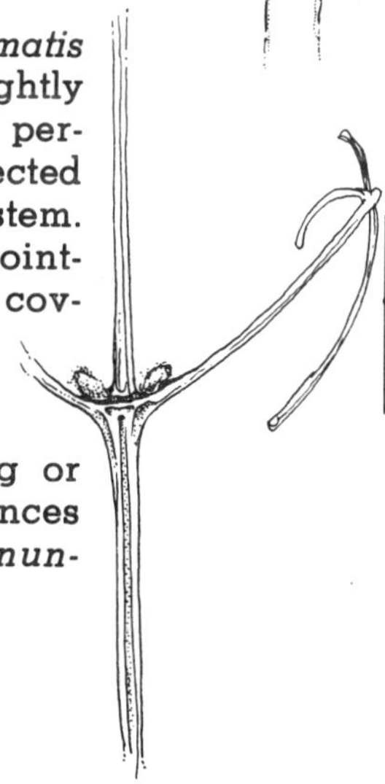

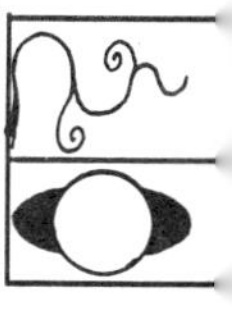

VIRGIN'S BOWER — *Clematis maximowicziana.* Lithe, slightly woody vines with opposite persistent leaf petioles connected by their bases around the stem. Imbricate, opposite buds, pointing out at a low angle, are covered with long, coarse, gray hairs. Fuzz buttons, terminating branches, are clusters of feathery fruits containing single seeds. Climbs by the bending or clasping of leaf stalks. Forms delicate arbors on fences and thickets in moist soil and partial sunlight. *Ranunculaceae.*

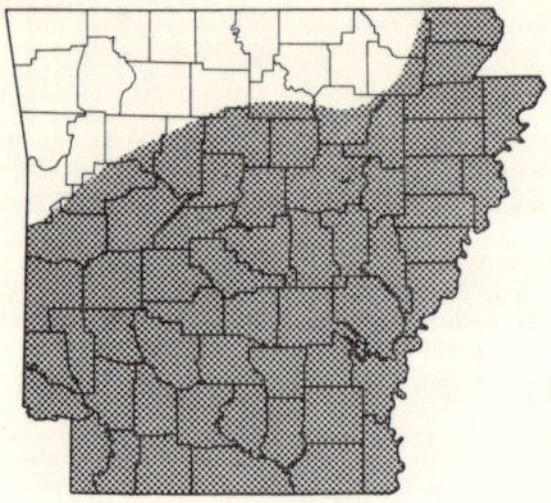

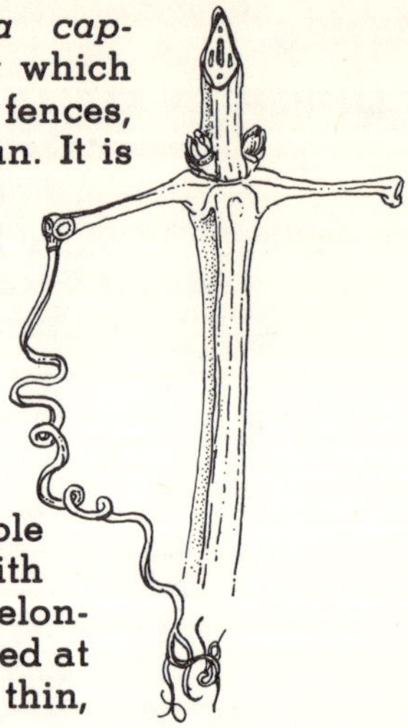

CROSS-VINE — *Bignonia capreolata.* A woody climber which forms dense arbors on fences, poles, and trees in full sun. It is reduced to a single, erect stem in shaded, moist woods. The four-angled stem has opposite, imbricate buds above persistent, perpendicular leaf petioles, suggesting a cross. Two leaflet scars and a branched tendril are at the end of the cross arms. Buds, young twigs, and some persistent leaves are strongly purple colored. Cross sections of the stem show spongy pith or none, as well as four isolated tissue areas. The elongated, dry fruits of thick parchment are long tapered at both ends and split into two halves, releasing many thin, winged seeds. *Bignoniaceae.*

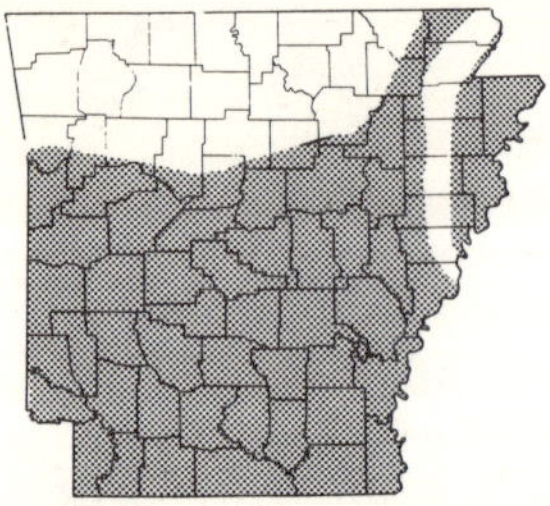

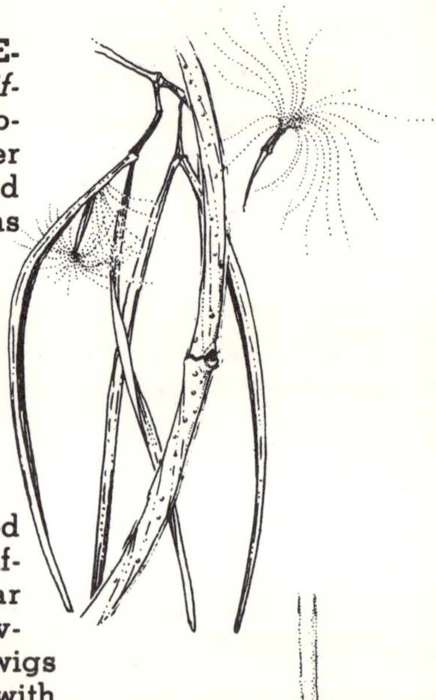

CLIMBING DOGBANE- *Trachelospermum difforme.* A thin vine climbing and twining over and around shrubs and trees at lake margins and wet ditches. Noticeable in winter by vast numbers of clusters of long, slender, drooping fruit follicles, splitting along two margins and twisting open. Long slender seeds are beaked at one end and tufted with many delicate hairs at the other. Fruit clusters arise from one of two opposite lateral buds. Leaf scars are opposite, elevated and joined by stipular scars. Leaf scar is half-round with a similar configuration of vascular bundle scars in the center. Lateral buds are covered by a single dark flap, fringed at the tip. Twigs smooth and light brown. Older vines gray with rusty brown, warty lenticels. *Apocynaceae.*

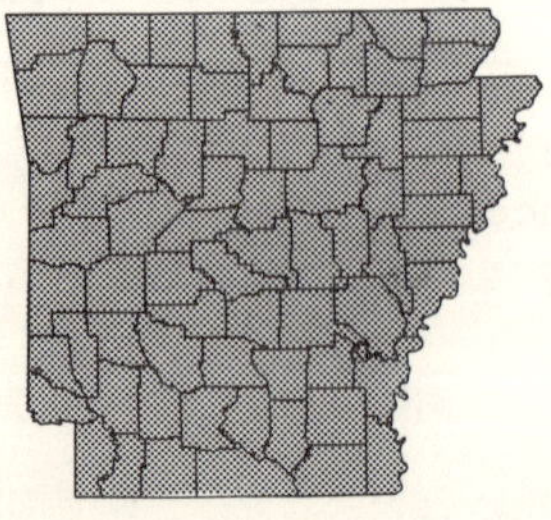

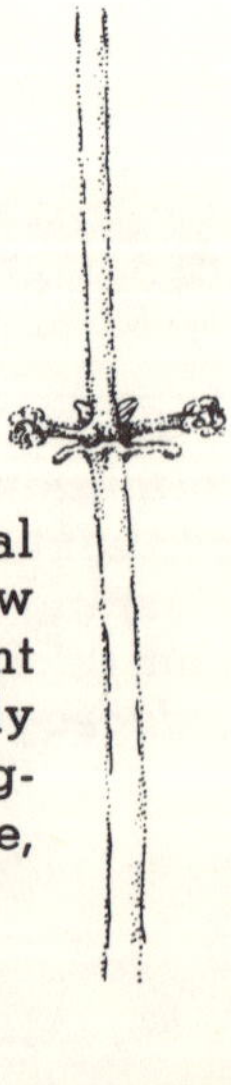

JAPANESE HONEYSUCKLE — *Lonicera japonica.* Long, slender, round, hairy vines have tip die-back. Lateral buds opposite, small, imbricated by a few hairy scales and obscured by persistent leaf petiole bases. No leaf scars. Mostly evergreen; sparsely deciduous. An aggressive, noxious vine forming dense, high, sprawling arbors and tangles. *Caprifoliaceae.*

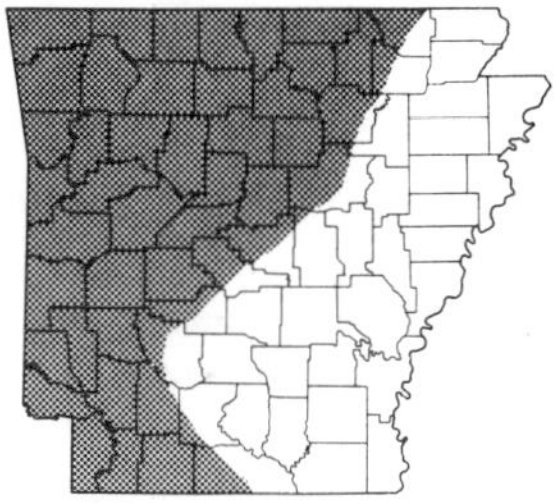

GRAPE HONEYSUCKLE — *Lonicera dioica.* Terminal bud usually abortive, resulting in forked and sparse branching by the opposite lateral buds. Buds loosely imbricated with chaffy scales. Bud scales and vining stems straw-colored, smooth, and dulled by black spatterings. Leaf petiole bases persistent or with chevron-shaped leaf scars. Stipular scars present. Vines are brittle, but tough, with a hollow pith. A sparse vine of talus slopes and dry sandstone ledges in upland hardwood forests. *Caprifoliaceae.*

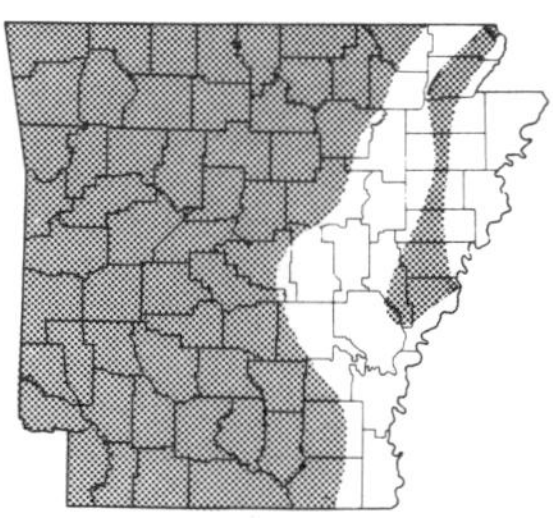

CORAL HONEYSUCKLE — *Lonicera sempervirens.* A delicate, smooth, straw-colored, brittle vine with opposite buds and die-back at the twig tips. Leaf scars are triangular to chevron-shaped, with three vascular bundle scars. Stipular scars form a connecting ridge between the leaf scars. Buds are loosely imbricated by papery, nearly transparent, gray scales. Native in moist open woods, but cultivated widely and escaping. *Caprifoliaceae.*

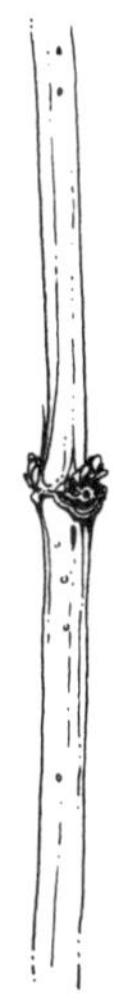

CLIMBING BY TENDRILS OPPOSITE THE BUDS

Summer Grape : Buds valvate. Scales reddish, membranous. Bark brown, shredding. p. 66

Muscadine : Twigs 5-angled, speckled white. Old stems smooth. p.67

Boston Ivy : Buds valvate. Leaf scar round. Holdfasts disc-tipped. Stem smooth, red. p. 67

Virginia Creeper : Buds imbricate. Leaf scar round. Holdfast disc-tipped. Stem warty, brown. p. 67

Pepper Vine : Buds imbricate : Leaf scar round. Tendrils slender-tipped. Black fruits. White continuous pith. p. 68

CLIMBING BY AXILLARY TENDRILS

Ladies' Eardrops : Buds imbricate, long-pointed. Stem orange, grooved. Tendrils coiled, copper-colored. p. 68

CLIMBING BY TWINING STEMS

Pipe Vine : Buds white, hairy, superposed. Twigs green, hairy, swollen at nodes. p. 68

Moonseed Vine : Buds superposed, the fourth in a crater. Stem green, smooth, ridged. p. 69

Carolina Moonseed : Buds superposed in a gray streak. Stem green, smooth, streaked. p. 69

Bittersweet : Buds short, sharp. Stipules withered. Twigs wiry-tipped. Stems twisted. Fruit orange and scarlet. p. 69

Wisterias : Buds valvate. Stipular scars present. Fruits knobby. p. 70

Rattan Vine : Buds minute. Twigs wiry. Stems red, then brown, then green. p. 70

Climbing Magnolia : Buds imbricate. Leaf scar ovate, raised. Old vines warty, aromatic. p. 71

CLIMBING BY AERIAL ROOTS

Poison Ivy : Buds naked, gray. Stems brown, dull, Fruit clusters axillary. Berries white. p. 71

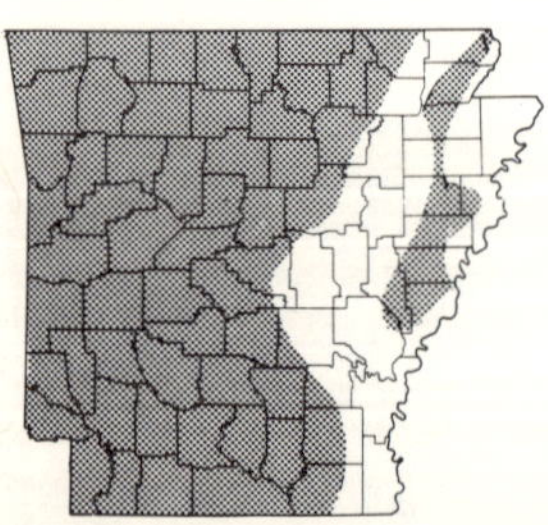

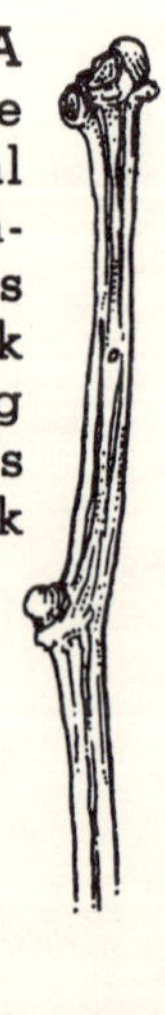

SUMMER GRAPE — *Vitis aestivalis.* A coarse vine climbing by tendrils that arise opposite some of the lateral buds. Lateral buds are valvate with reddish membranous scales. Round to half-round leaf scars have stipular scars at the side. Old bark loosens and shreds off. Old, high-climbing vines reach their maximum size in uplands and high floodplains forests. Gray back grape, *V. cinerea,* shares essentially the same habitat, but *V. palmata,* red grape, and *V. riparia,* riverside grape, are encountered only in low floodplains, bayous, and swamps. *Vitaceae.*

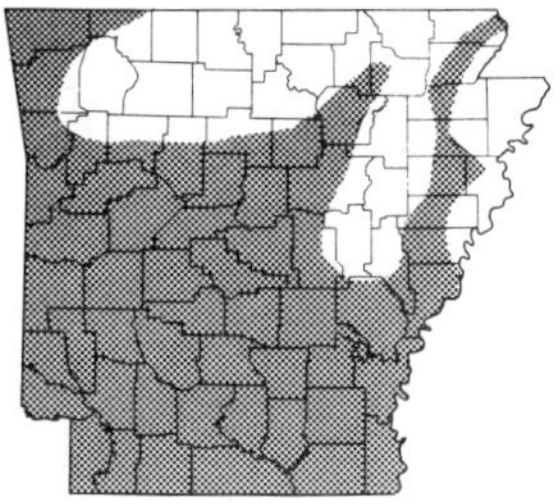

MUSCADINE — *Vitis rotundifolia.* A running, climbing, or sprawling, dark, smooth vine forms a tight, stiff arbor. Blunt twigs show a flat, broad terminal bud scar next to the pseudo-terminal bud and leaf scar. Alternate, oblique, lateral buds have smooth, membranous, valvate scales which give way to reddish-brown, hairy inner scales. Leaf scars are half-elliptical to triangular and ringed by vascular bundle scars near the outer edge. Leaf scars are raised at swollen nodes and attended by thin stipular scars atop ridges that turn back down the stem. Twigs are faintly five-angled with numerous, fine, white lenticels that speckle the dark twigs. Tendrils arise opposite the lateral buds infrequently, and bind the stem to a support or to itself. *Vitaceae.*

BOSTON IVY — *Parthenocissus tricuspidata.* Twigs show die-back or a terminal bud scar accompanied by a pseudo-terminal bud. Lateral buds occur with a frequency of about every third or fourth node. The presence of reduced, branched, disc-tipped holdfasts correlates with no lateral bud at the leaf scar. The twigs are rough because of round, raised leaf scars, stipular scars, projecting buds, holdfasts, and elongated lenticels. Otherwise, the vine is glabrous and dark reddish-brown. Introduced from China and Japan for ornamental climbers. Usually found on urban buildings. *Vitaceae.*

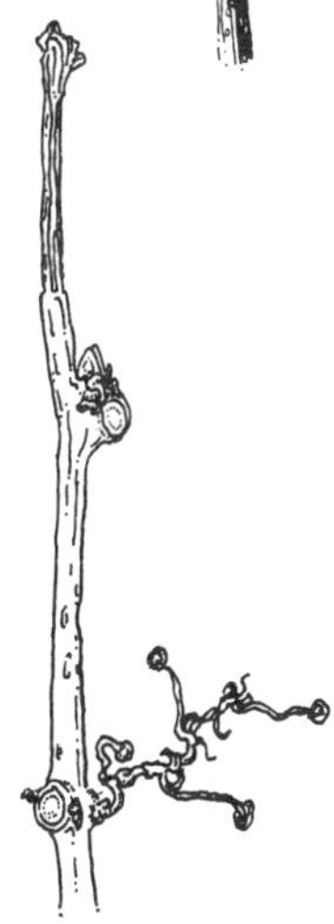

VIRGINIA CREEPER — *Parthenocissus quinquefolia.* Twigs show die-back or terminal bud scars. Lateral buds alternate, short-conical, imbricate, light brown, and dry. The rimmed, saucer-shaped, leaf scars are raised, giving the twig a strong zig-zagged and rough profile; stipular scars present. Branched holdfasts, with adhesive pads, arise opposite the earlier leaf scars. Leaf scars across from branching holdfasts have no lateral bud. Produces a forest floor cover in reduced light or becomes an aggressive, high climber along open floodplains. *Vitaceae.*

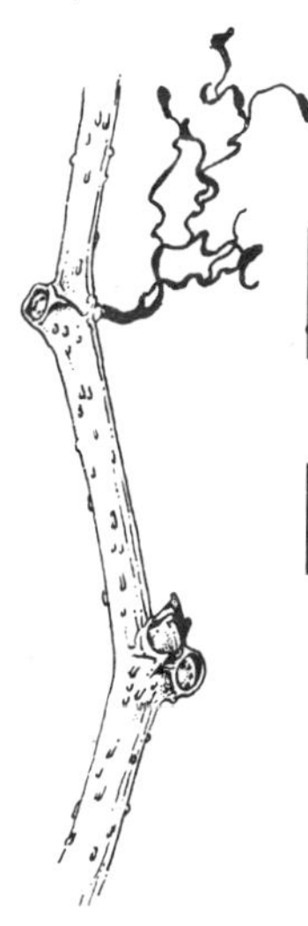

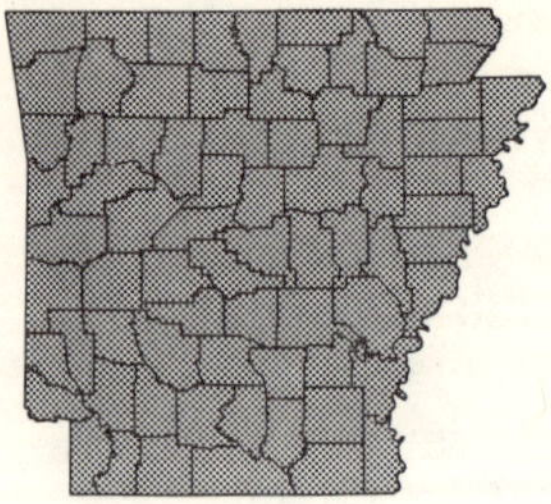

PEPPER VINE — *Ampelopsis arborea.* Pseudo-terminal buds are accompanied by a round or half-round and raised leaf scar. Stipular scars present. The light gray-brown bark supports many lenticels. No nodal partition interrupts the continuous white pith. This sprawling, shrubby vine climbs by forked, slender-tipped tendrils. Clusters of black fruits persist into late fall. It forms low arbors in low woods and swampy bottomland. Cultivated widely. False Grape, *A. cordata,* resembles peppervine closely in its winter condition, differing by its bluish-purple to green fruits and by a chambered pith. *Vitaceae.*

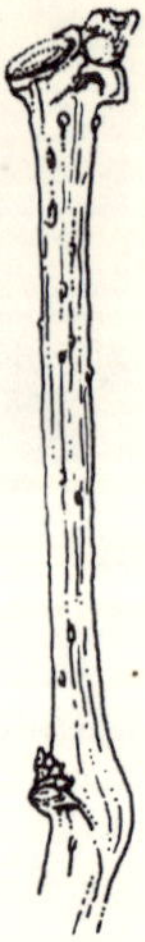

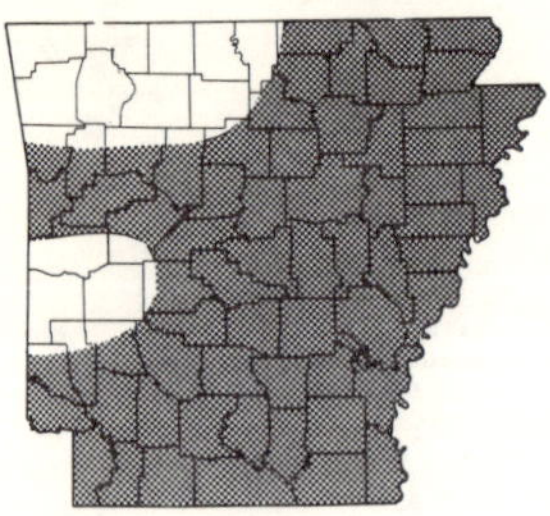

LADIES' EAR-DROPS — *Brunnichia ovata.* A delicate, wiry vine, climbing on trees of streambanks. The orange, grooved stems have right-angled branches terminated by tightly coiled, copper-colored tendrils. Twigs are deeply grooved and twisted; die-back occurs at the tips. Lateral buds are alternate, imbricate, long-pointed, compressed, and tongue-like. The bud is between a branch above, and the leaf petiole remnant below. Leaf scars absent. The persistent petiole stub is dilated and continues around the twig as a narrow membraneous stipule. *Polygonaceae.*

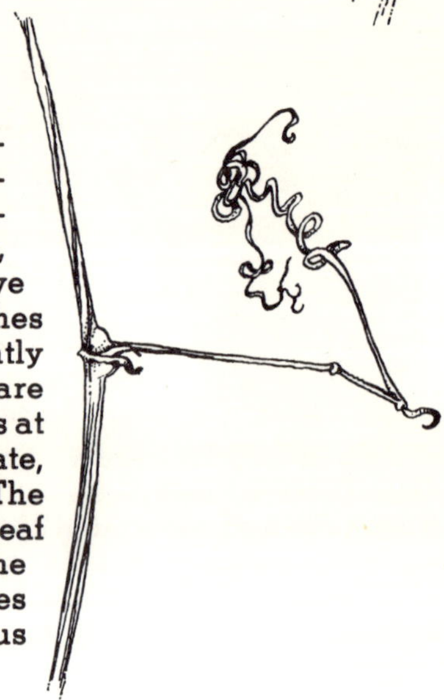

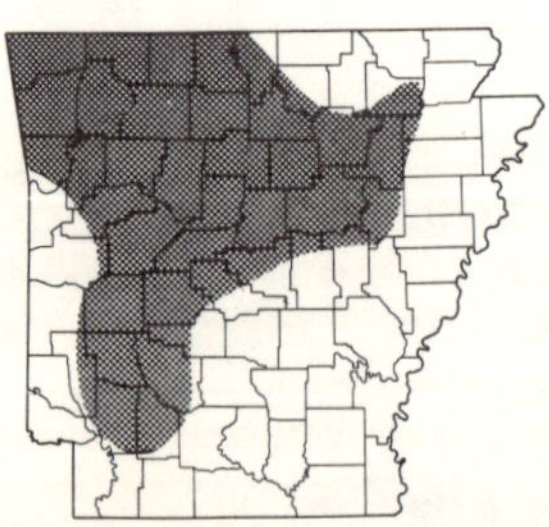

PIPE VINE — *Aristolochia tomentosa.* A high-twining or shrubby vine with gray, corky ridged, and old vines that entwine, often forming a cable. Younger stems, dark with elongate, erupting lines of cork that resemble thin, pursed lips. These vines crunch when bent. Twig is bright green and hairy, noticeably swollen at the nodes. Three white, hairy buds, (one large, two small) are on a velvety pad inside the thin, U-shaped leaf scar. Leaf scars are black with three, white vascular bundle scars. Blunt twigs are caused by a broad, flat, oblique, terminal bud scar. Found in rich, wooded alluvial or calcareous soils. *Aristolochiaceae.*

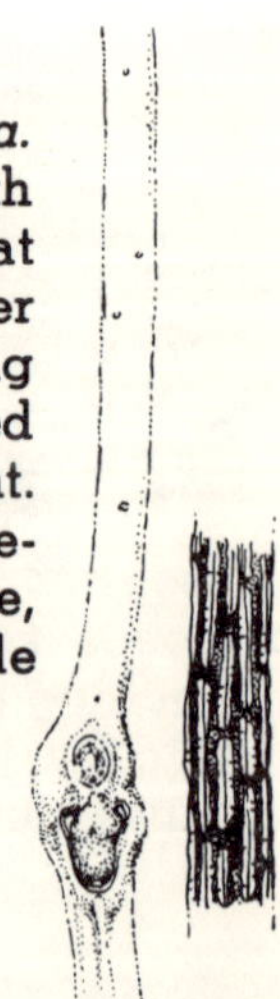

MOONSEED — *Menispermum canadense.* A smooth, shiny green, twisting, finely ridged, flexible vine with tip die-back. Leaf scars are alternate, raised, and elliptical, splitting at the top by an emerging bud. Four buds are superposed above the leaf scar, the fourth and largest, in a crater. The vines are pithy and a cross section shows the wood separated by wide rays into at least twenty sections. Rich thickets and streambanks. *Menispermaceae.*

CAROLINA MOONSEED — *Cocculus carolinus.* A high climbing or long running smooth, green vine with fine lines running its length. Sparsely branched and with tip die-back. Round leaf scars are alternate, slightly rimmed, inconspicuous. A gray, hairy streak above each leaf scar hides three superposed buds. A cross section of the stem shows a pithy center and wood divided at least twenty times by thin pith rays. Rich woods, thickets, and stream margins support its growth in thin arbors. *Menispermaceae.*

AMERICAN BITTERSWEET — *Celastrus scandens.* A twining, sprawling, arbor-forming vine producing terminal clusters of ornamental fruits in late fall. The orange fruit coats split and expose the scarlet seeds. Twigs attenuated into wiry curls. Buds alternate and imbricated by short, sharp, erect bud scales. Leaf scars elevated, small, round, attended by minute, withered stipules, and show a central patch of vascular bundle scars. Stems twisted, angular, light-weight, and dull gray-brown. Pith randomly splotched with orange. Roadsides, fencerows, and open dry woods. *Celastraceae.*

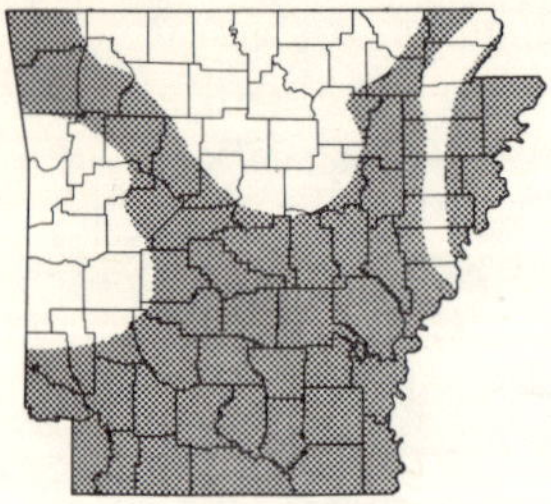

WISTERIA — *Wisteria frutescens.* A coarse, shrubby or high-climbing vine of streambanks in low, wet woods. Conpicuous in winter by clusters of smooth, dull, reddish-brown, knobby pods on tan twigs. Older vines are olive speckled with light, corky lenticels. Twigs hirsute, hard but pithy; blunt and swollen by a terminal bud scar, leaf scar, and the attending lateral bud. Buds valvate with tomentose scales. Leaf scars alternate, half-round, raised, swollen, and showing stipular scars. Found with tupelo gum, cypress, overcup oak, and planer tree. *Leguminosae.*

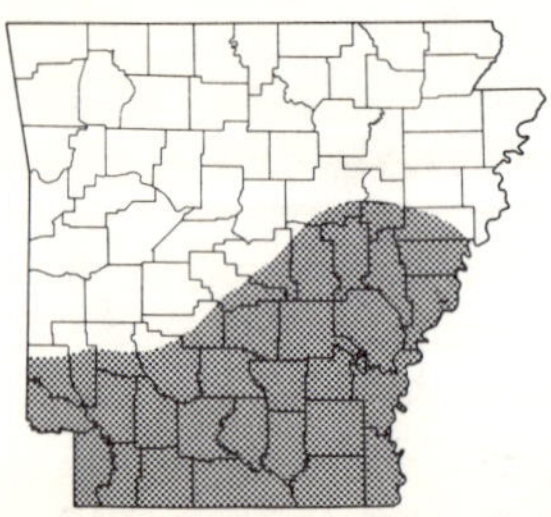

WISTERIA — *Wisteria floribunda.* A coarse, vigorous, high-climbing and twining vine. Its fruits are long, velvety, knobby pods. The twigs and older stems are rough because of raised, protruding, and rounded leaf scars. The valvate, lateral buds are mostly alternate but some are subopposite. Cultivated extensively at house sites and public places. *Leguminosae.*

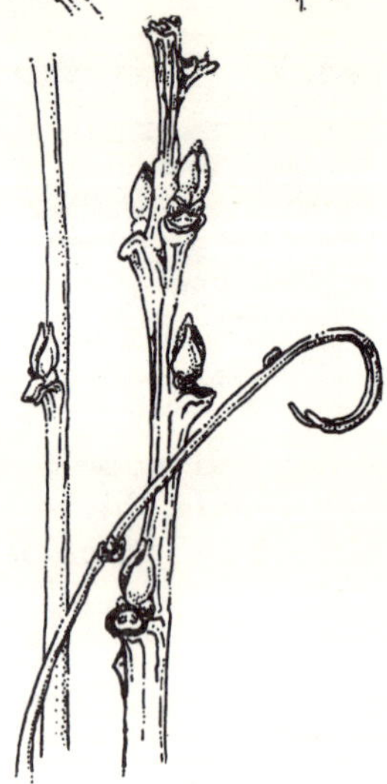

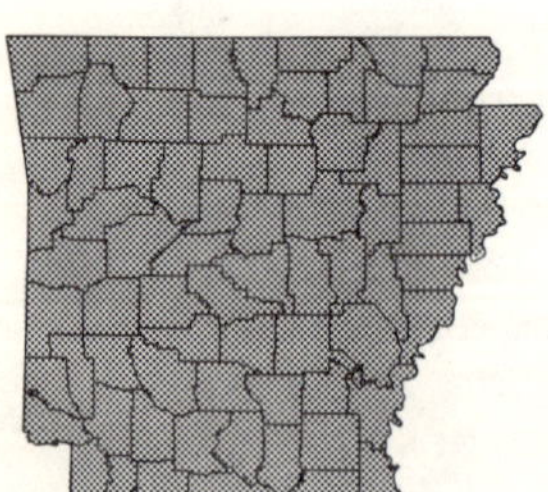

RATTAN VINE — *Berchemia scandens.* A hard, smooth, hightwining vine. Twigs red, then brown, then green, with age. Fruits abundant, small, and blue, with bony center; clustered in small panicles. Twigs wiry and irregular with raised leaf scars and minute buds. The "supple-jack" of all wooded habitats of the state. *Rhamnaceae.*

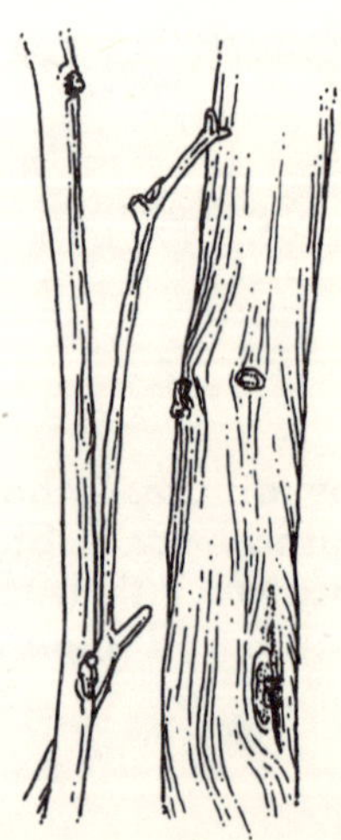

CLIMBING MAGNOLIA — *Schisandra glabra.* A limber, sparsely branching vine growing as a ground cover or as a low twiner in mixed hardwood forests of loessial soils. Frequent seasonal die-back forces a pattern of lateral branching. Pseudo-terminal bud slightly stalked and pointed, accompanied by a terminal bud scar or a black withered stalk. Buds are alternate, imbricated by reddish-brown, membraneous, ciliate scales, and point sharply outward from the dull gray-brown twig. Leaf scars elevated, generally ovate with three vascular bundle scars arranged as a central chevron or arc. Old vines become warty. Bruised bark is faintly aromatic. Restricted in range to Crowley's Ridge below Poinsett County. *Schisandraceae.*

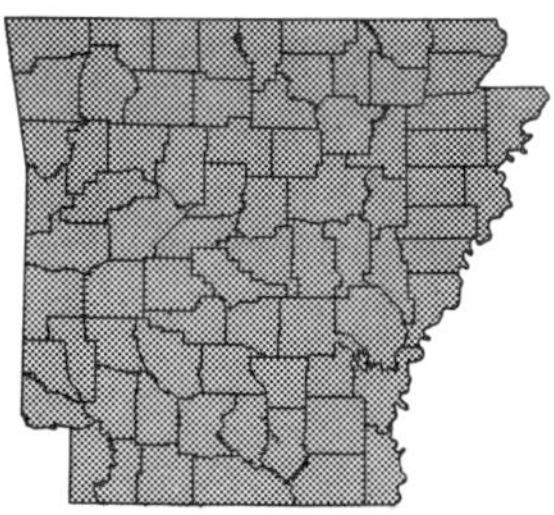

POISON IVY — *Toxicodendron radicans.* Erect, low shrubs or coarse vines that climb and attach to a support by vast numbers of aerial roots. Twigs appear dull, light brown with gray, naked, hairy buds. Buds alternate in the notch of a broadly crescent leaf scar. Later fruit panicles bear white or cream-colored berries. A ubiquitous and toxic vine. *Anacardiaceae.*

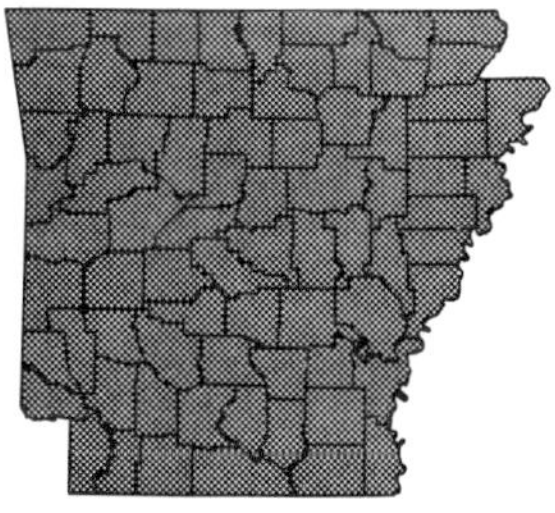

GREENBRIER — *Smilax rotundifolia.* A hard, coarse, dark-green, sprawling, vine armed with prickles and spines, flattened at the base. Leaf scar covered by the withered leaf petiole base. Leaf scar often attended by two stipular spines. *Liliaceae.*

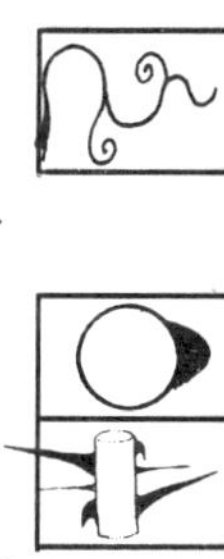

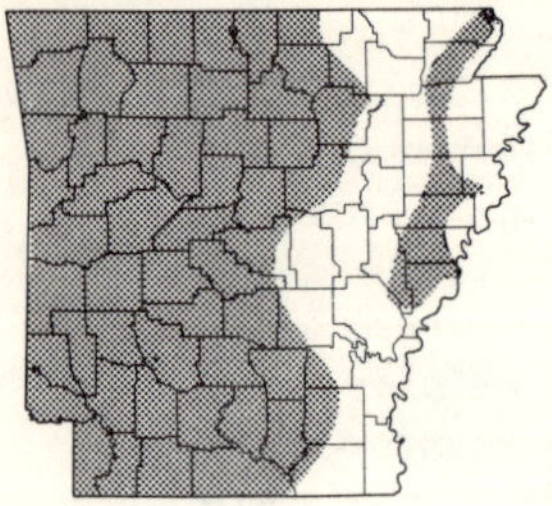

SAWBRIER — *Smilax glauca.* A climbing, entangled arbor-forming vine, appearing fine-textured and dull. Twigs zigzagged, wiry, smooth, shiny red-speckled. Persistent stipular tendrils coil back and around the stem. Alternate lateral buds are covered by old stub of the leaf petiole. Short rigid spines arise laterally from each side of older lateral buds. Fleshy fruit, five or six in a round cluster, borne laterally on new wood. Wide spread but shows optimal growth in wet woods and lake margins. *Liliaceae.*

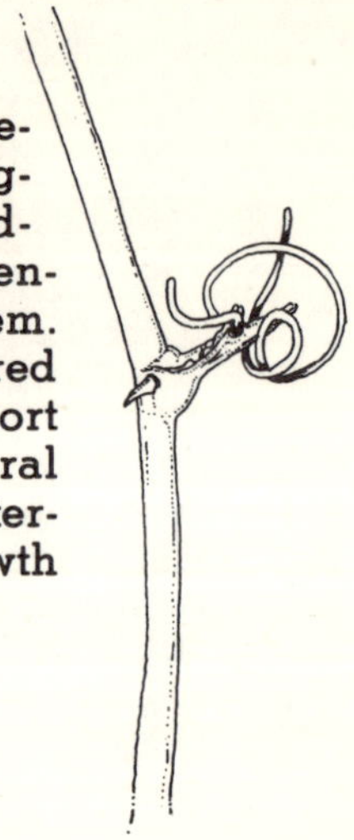

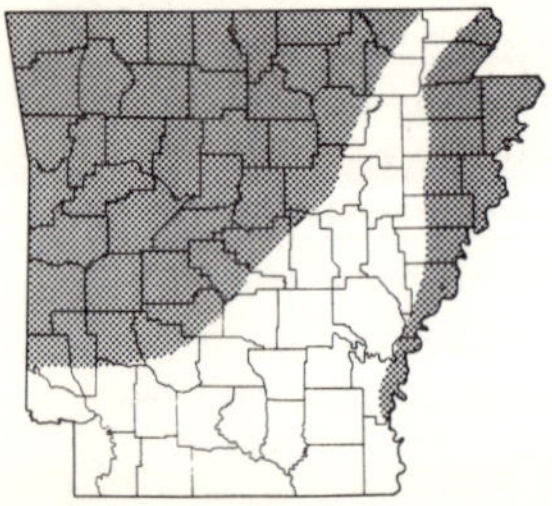

BRISTLY GREENBRIER — *Smilax tamnoides.* A hard, green, high climbing vine with numerous, thin, black, flexible prickles flattened at the base end of variable size. Old stems are either very bristly or slightly bristly; sometimes bristles are absent on twigs. Alternate lateral buds partially covered by persistent leaf petiole base. Single fruits are borne on a thin, flexible stalk arising out of the axillary bud on new wood. Vines without pith. Found in low woods and thickets. *Liliaceae.*

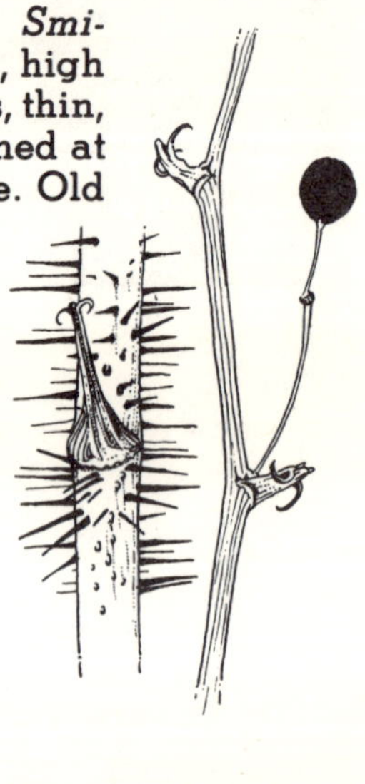

Glossary

ACCESSORY BUDS : Multiple lateral buds per leaf scar. 

ACUTE : Sharp. An angle of ninety degrees or less.

ADVENTITIOUS BUD : A bud arising out of the ordinary place. 

ALTERNATE BUDS : Buds occurring on first one side and then the other at different nodes. 

AMENTS : Dense elongated cylinders of bracted flower-bud clusters.

APPARENT BUDS : Obvious; not embedded.

APPRESSED : Pressed close to.

BRACTS : Other than ordinary leaves; specialized leaves.

BRANCH SCAR : A scar, without a bud, left when a branch is lost.

BRISTLY : Covered with stiff, strong hairs. 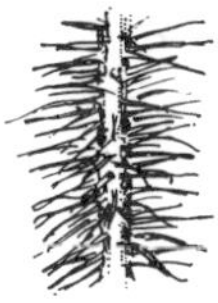

BUD : An undeveloped branch enclosing immature leaves, flowers, or both; may or may not be enclosed by scales. 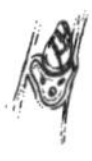

CHAMBERED : Air space interrupted by walls.

CILIATE : With conspicuous marginal hairs on the bud scale. 

COLLATERAL BUDS : Accessory buds that occur on either side of the normal lateral buds. 

CONICATE BUDS : Cone-shaped by a single scale. 

CONVOLUTED : Twisted.

COPPICED : The effect of repeated pruning on a tree: a terminal thicket of even-aged branches.

DECIDUOUS : Having a complete loss of annual leaves.

DECURRENT : Extending downward.

DENTICULAR : Tooth-like.

DIE-BACK : Late growth of a twig killed back by low temperatures.

DOWNY : Covered with short, weak, soft hairs.

DULL : Not glossy; lacks lustre.

ELEVATED : Raised in position.

EMBEDDED BUD : Main body of bud is below the surface of the leaf scar or twig.

EXCURRENT : With a single, undivided, main trunk.

FERAL : Undomesticated, gone wild.

FLOCCOSE : Covered with uneven patches or tufts of dense, matted hairs.

FLOWER BUDS : Enlarged buds which produce only flowers.

FLOWER/FRUIT SCARS : Similar in appearance to branch scars; more often found in a terminal position.

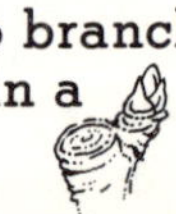

FLUTED : Bark with erect ruffles.

GLABROUS : Smooth, not hairy.

GLAUCOUS : Having a powdery or waxy coating that can be rubbed off easily.

GLOBOSE : As a globe, spherical.

HIRSUTE : With sparse, long, hairs.

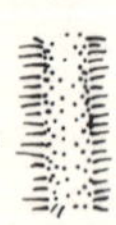

IMBRICATE BUDS : Buds with scales overlapped in regular order.

INTERNODES : In between the nodes.

INVOLUCRE : A whorl of enclosing or subtending bracts.

KEELED : With a central ridge down the center of the scale.

LANCEOLATE : Lance-shaped, gradually tapered to a point.

LATERAL BUD : A side bud with a leaf scar. Located other than at the twig tip.

LEAF SCAR : The scar left when a leaf falls off.

LENTICELS : Small corky eruptions on woody stems. 

LUSTROUS : Reflecting light. Shining.

MEMBRANOUS : A thin parchment-bordered bud scale. 

NAKED BUD : A bud that lacks bud scales. May have a coating of resin, hairs, or both. 

NODE : Regions on stems enlarged by buds and leaf scars.

OBLIQUE : Slanting away from the center line of the twig. 

OBTUSE : Blunt. An angle greater than ninety degrees.

OPPOSITE BUDS : Two buds, each with a separate leaf scar, on opposing sides of the stem. 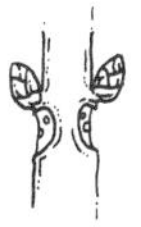

PANICLE : A diffuse and loosely branched cluster of fruits.

PETIOLE : The leaf stalk. 

PRICKLES : Short, sharp-pointed projections arising from the surface layers of a twig. Superficial. Can be removed. 

PSEUDO-TERMINAL BUD : A lateral bud which takes a tip position in the absence of a terminal bud. 

PUBESCENT : Covered by straight slender hairs. 

RIDGE : A linear projection above the surface of a stem. 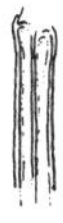

SCABROUS : Rough to the touch; sand-papery. 

SCURFY : With small, dry, loose particles.

SHRUB : A low, woody plant with erect, multiple stems. A bush.

SPINE : A sharp woody outgrowth from the position of a leaf or leaf part. 

SPLAYED : Spread out at a low angle.

SPUR : A fruit spur. A branch dwarfed by repeated fruit production.

STIPULAR SCARS : Lateral extensions of the leaf scar caused by loss of a stipule.

STIPULAR SPINES : Spines formed by dry, hardened stipules.

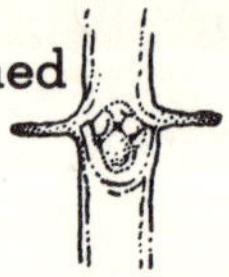

STIPULES : A pair of membraneous leaf appendages that arise from the base of the leaf of some plants.

STRIATED : Marked with narrow bands of parallel lines.

SUB-OPPOSITE : Nearly opposite.

SUBTEND : To extend under.

SUPERPOSED BUDS : Accessory buds that occur above the normal lateral bud.

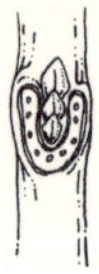

SUPPLE : Flexible.

TENDRIL : A slender, coiling organ of climbing and attachment function.

TERMINAL BUD : A bud at the tip end of a twig. It has no leaf scar.

TERMINAL BUD SCALE SCARS : A series of scars arranged in a band around the twig.

TERMINAL BUD SCAR : Scar left when the terminal bud is lost.

THORN : A short, hard, sharp-pointed, and usually leafless branch.

TOMENTOSE : Densely pubescent with matted wool.

TOMENTUM : A tomentose covering.

TREE : A woody perennial plant having a single erect, basal trunk. A small tree usually does not exceed fifteen feet in height.

VALVATE BUDS : Buds in which two scale edges meet without overlap. Often with more than one set of scales.

VASCULAR BUNDLE SCARS : One to many small dots or lines found on the surface of the leaf scar. 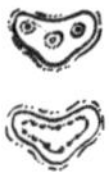

VEGETATIVE BUD : A common-bud that develops into a leafy shoot only.

VINE : Any plant with a weak, sprawling, vining, climbing stem.

WATER-SPROUT : An erect, adventitious, vigorous stem arising from an arching limb.

WINGS : Flat, thin, corky expansions on stems. 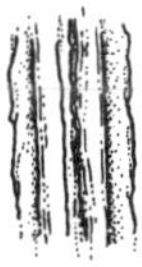

WHORLED BUDS : An arrangement of three or more buds at a node, each with a leaf scar. 

GEOMETRIC FIGURES:

ELLIPTICAL

HALF-ELLIPTICAL

HALF-ROUND

OBOVATE

OVATE

CHEVRON-SHAPED

OBTRIANGULAR

RHOMBIC

TERETE

RENIFORM

CRESCENT

References

BAILEY, L.H. & E.S. 1976. Hortus Third. Macmillan Co., Inc. N.Y. 1290 p.

FERNALD, M.L. 1950. Gray's Manual of Botany. 8th Edition. American Book Co., N.Y. 1632 p.

FOTI, THOMAS. 1974. Natural Divisions of Arkansas, *in* Arkansas Natural Area Plan. Arkansas Department of Planning.

HARLOW, WILLIAM M. 1959. Fruit Key and Twig Key to Trees and Shrubs. Dover Publications, Inc. N.Y. 56 p.

MOORE, D.M. 1972. Trees of Arkansas. Arkansas Forestry Commission. Little Rock. 142 p.

PELL, W.F., JR. 1980. Vegetation Types of Arkansas: A Working Scheme for Identification and Classification. Unpublished. Arkansas Heritage Program, Nature Conservancy. Little Rock. 11 p.

PRESTON, R.J., JR. 1976. North American Trees. 3rd Edition. The Iowa State University Press, Ames, Iowa. 399 p.

SMITH, EDWIN B. 1978. An Atlas and Annotated List of Vascular Plants of Arkansas. 592 p.

STEYERMARK, JULIAN A. 1963. Flora of Missouri. The Iowa State University Press, Ames, Iowa. 1728 p.

TUCKER, GARY. 1975. A Guide to the Woody Flora of Arkansas. Unpublished Dissertation. University of Arkansas. 207 p.

Families of Deciduous Woody Plants of Arkansas

The asterisk (*) marks the species content of the book.

CLASS GYMNOSPERMAE

TAXODIACEAE — Bald Cypress Family
* *Taxodium distichum* — Bald Cypress

GINKGOACEAE — Ginkgo Family
* *Ginkgo* — Ginkgo biloba

CLASS ANGIOSPERMAE
SUBCLASS DICOTYLEDONAE

ACERACEAE — Maple Family
* *Acer negundo* — Box Elder
* *A. rubrum* — Red Maple
  A. rubrum var. *drummondii* — Swamp Maple
* *A. saccharinum* — Silver Maple
* *A. saccharum* — Sugar Maple
  A. saccharum var. *floridanum* — Southern Sugar Maple
  A. saccharum var. *leucoderme* — Chalk Maple
  A. saccharum var. *nigrum* — Black Maple

ANACARDIACEAE — Chashew Family
* *Cotinus obovatus* — American Smoke Tree
* *Rhus aromatica* — Aromatic Sumac
  R. aromatica var. *flabelliformis* — Aromatic Sumac
  R. aromatica var. *serotina* — Aromatic Sumac
* *R. copallina* — Winged Sumac
* *R. glabra* — Smooth Sumac
* *Toxicodendron radicans* — Poison Ivy
  T. toxicarium — Poison Oak

ANNONACEAE — Custard Apple Family
  Asimina parviflora — Dwarf Pawpaw
* *A. triloba* — Pawpaw

APOCYNACEAE — Dogbane Family
* *Trachelospermum difforme* — Climbing Dogbane

AQUIFOLIACEAE — Holly Family
* *Ilex decidua* — Possum Haw
  I. decidua var. *longipipes* — Possum Haw
* *I. verticillata* — Black Alder

ARALIACEAE — Ginseng Family
* *Aralia spinosa* — Hercules' Club

ARISTOLOCHIACEAE — Birthwort Family
* *Aristolochia tomentosa* — Pipe Vine

BETULACEAE — Birch Family
* *Alnus serrulata* — Common Alder
* *Betula nigra* — River Birch
* *Carpinus caroliniana* — American Hornbeam
* *Corylus americana* — Hazelnut
* *Ostrya virginiana* — Hop Hornbeam

BIGNONIACEAE — Trumpet Creeper Family
* *Bignonia capreolata* — Cross-vine
* *Campsis radicans* — Trumpet Creeper
 Catalpa bignoniodes — Southern Catalpa
* *C. speciosa* — Northern Catalpa

CAPRIFOLIACEAE — Honeysuckle Family
* *Lonicera dioica* — Grape Honeysuckle
* *L. japonica* — Japanese Honeysuckle
 L. mackii — Honeysuckle
* *L. sempervirens* — Coral Honeysuckle
* *Sambucus canadensis* — Common Elderberry
* *Symphoricarpos orbiculatus* — Coral Berry
* *Viburnum dentatum* — Southern Arrow-wood
 V. molle — Arrow-wood
* *V. nudum* — Shonny Haw
 V. prunifolium — Black Haw
 V. rafinesquianum — Downy Arrow-wood
 V. recognitum — Arrow-wood
* *V. rufidulum* — Rusty Black Haw

CELASTRACEAE — Staff-tree Family
 Celastrus orbiculatus — Oriental Bittersweet
* *C. scandens* — American Bittersweet
* *Euonymus americanus* — Strawberry Bush
* *E. atropurpureus* — Wahoo
 E. obovatus — Running Strawberry Bush

COMPOSITAE — Composite Family
* *Baccharis halimifolia* — Sea-Myrtle

CORNACEAE — Dogwood Family
 Cornus alternifolia — Alternate-leaved Dogwood
* *C. amomum ssp. obliqua* — Silky Dogwood
* *C. drummondii* — Roughleaf Dogwood
* *C. florida* — Flowering Dogwood
* *C. foemina* — Stiff Dogwood

EBENACEAE — Ebony Family
* *Diospyros virginiana* — Persimmon

ERICACEAE — Heath Family
* *Gaylussacia baccata* — Black Huckleberry
* *Lyonia ligustrina* — Male-berry
* *L. mariana* — Staggerbush
* *Rhododendron canescens* — Swamp Azalea
* *R. oblongifolium* — White Azalea
* *R. prinophyllum* — Election Pink
* *Vaccinium arboreum* — Farkleberry
* *V. elliotii* — Mayberry
* *V. fuscatum* — Dusky Blueberry

* *V. pallidum* — Low-bush Huckleberry
* *V. stamineum* — Deerberry
V. virgatum

EUPHORBIACEAE — Spurge Family
Andracne phyllanthoides — Maidenbush
* *Sapium sebiferum* — Chinese Tallow Tree

FAGACEAE — Beech Family
* *Castanea pumila* — Downy Chinquapin
* *C. pumila* var. *ozarkensis* — Ozark Chinquapin
* *Fagus grandifolia* — Beech
* *Quercus alba* — White Oak
* *Q. arkansana* — Arkansas Oak
* *Q. coccinea* — Scarlet Oak
Q. durandii — Durand's White Oak
* *Q. falcata* — Southern Red Oak
Q. falcata var. *pagodaefolia* — Cherrybark Oak
* *Q. imbricaria* — Shingle Oak
Q. incana — Bluejack Oak
* *Q. lyrata* — Overcup Oak
* *Q. macrocarpa* — Bur Oak
* *Q. marilandica* — Blackjack Oak
* *Q. michauxii* — Cow Oak
* *Q. nigra* — Water Oak
* *Q. nuttallii* — Nuttall's Oak
* *Q. palustris* — Pin Oak
* *Q. phellos* — Willow Oak
* *Q. prinoides* var. *acuminata* — Chinkapin Oak
Q. prinoides var. *prinoides* — Dwarf Chinkapin Oak
* *Q. rubra* — Northern Red Oak
* *Q. shumardii* — Shumard's Oak
* *Q. shumardii* var. *acerifolia* — Maple-leaved Oak
* *Q. stellata* — Post Oak
Q. stellata var. *margaretta* — Margaretta Oak
Q. stellata var. *mississippiensis* — Delta Post Oak
* *Q. velutina* — Black Oak

HAMAMELIDACEAE — Witch Hazel Family
* *Hamamelis vernalis* — Vernal Witch Hazel
* *H. virginiana* — Eastern Witch Hazel
* *Liquidambar styraciflua* — Sweet Gum

HIPPOCASTANACEAE — Horse Chestnut Family
* *Aesculus glabra* — Ohio Buckeye
* *A. pavia* — Red Buckeye

HYPERICACEAE — St. John's-wort Family
Hypericum densiflorum var. *lobocarpum* — St. John's Wort
H. hypericoides — St. Andrew's Cross
H. hypericoides var *multicaule*
H. nudiflorum
H. prolificum — Shrubby St. John's Wort
H. stans — St. Peter's Wort

JUGLANDACEAE — Walnut Family
* *Carya aquatica* — Water Hickory
* *C. cordiformis* — Bitternut Hickory
C. glabra — Pignut Hickory

* *C. illinoensis* — Pecan
C. lacinoiosa — Shellbark Hickory
C. myristiciformis — Nutmeg Hickory
C. ovalis — Red Hickory
* *C. ovata* — Shagbark Hickory
* *C. texana* — Black Hickory
* *C. tomentosa* — Mockernut Hickory
* *Juglans cinerea* — Butternut
* *J. nigra* — Black Walnut

LAURACEAE — Laurel Family
* *Lindera benzoin* — Spicebush
L. melissifolia — Pond Berry
* *Sassafras albidum* — Sassafras

LEGUMINOSAE — Legume Family
* *Albizia julibrissin* — Mimosa
Amorpha canescens — Lead Plant
* *A. fruticosa* — Indigo Plant
A. nitens
A. ouachitensis
A. paniculata
* *Cercis canadensis* — Redbud
* *Cladrastis kentudea* — Yellow Wood
* *Gleditsia aquatica* — Water Locust
* *G. triacanthos* — Honey Locust
* *Gymnocladus dioicus* — Kentucky Coffee Tree
* *Robinia hispida* — Bristly Locust
* *R. pseudo-acacia* — Black Locust
Sophora affinis — Texas Sophora
* *Wisteria floribunda* — Wisteria
* *W. frutescens* — Wisteria
W. sinensis — Chinese Wisteria

LEITNERIACEAE — Corkwood Family
* *Leitneria floridana* — Corkwood

MAGNOLIACEAE — Magnolia Family
* *Liriodendron tulipifera* — Tulip Tree
* *Magnolia acuminata* — Cucumber Tree
* *M. soulangiana* — Soulange's Magnolia
* *M. tripetala* — Umbrella Magnolia

MALVACEAE — Mallow Family
* *Hibiscus syriacus* — Rose-of-Sharon

MELIACEAE — Mahogany Family
* *Melia azedarach* — Chinaberry

MENISPERMACEAE — Moonseed Family
Calycocarpum lyoni — Cupseed
* *Cocculus carolinus* — Carolina Moonseed
* *Menispermum canadense* — Moonseed

MORACEAE — Mulberry Family
Broussonetia papyrifera — Paper Mulberry
* *Maclura pomifera* — Bois D'Arc
Morus alba — White Mulberry
* *M. rubra* — Red Mulberry

NYSSACEAE — Tupelo Family
* *Nyssa aquatica* — Tupelo Gum
* *N. sylvatica* — Black Gum
N. sylvatica var biflora — Swamp Black gum

OLEACEAE — Olive Family
* *Chionanthus virginicus* — Fringe Tree
* *Forestiera acuminata* — Swamp Privet
* *Fraxinus americana* — White Ash
F. caroliniana — Water Ash
* *F. pennsylvanca* — Green Ash
* *F. profunda* — Pumpkin Ash
* *F. quadrangulata* — Blue Ash

PLANTAGINACEAE — Plane Tree Family
* *Platanus occidentalis* — Sycamore

POLYGONACEAE — Buckwheat Family
* *Brunnichia ovata* — Ladies' Eardrops

RANUNCULACEAE — Crowfoot Family
* *Clematis maximowicziana* — Virgin's Bower
C. virginiana — Virgin's Bower

RHAMNACEAE — Buckthorn Famly
* *Berchemia scandens* — Rattan Vine
* *Ceanothus americanus* — New Jersey Tea
C. herbaceus — Redroot
* *Rhamnus caroliniana* — Carolina Buckthorn
* *R. lanceolata* — Buckthorn

ROSACEAE — Rose Family
* *Amelanchier arborea* — Shadbush
* *Crataegus crus-galli* — Cockspur Thorn
(See Smith and Tucker for complete treatment of 16-18 species).
* *Neviusia alabamensis* — Snow Wreath
* *Physocarpus opulifolius* — Ninebark
* *Prunus americana* — Wild Plum
P. angustifolia — Chickasaw Plum
P. hortulana — Hortulan Plum
P. mahaleb — Perfumed Cherry
* *P. mexicana* — Big Tree Plum
P. munsoniana — Wild Goose Plum
* *P. persica* — Peach
P. pumila — Sand Cherry
P. rivularis — Creek Plum
* *P. serotina* — Wild Black Cherry
P. umbellata — Flatwood Plum
P. virginiana — Choke Cherry
Pyrus angustifolia — Narrow-leaved Crab Apple
* *P. arbutifolia* — Red Chokeberry
* *P. communis* — Common Pear
P. coronaria — Wild Crab
* *P. floribunda* — Flowering Crab
P. ioensis — Crab Apple
* *P. malus* — Apple
P. melanocarpa — Black Chokeberry
Rosa arkansana — Sunshine Rose
R. bracteata — McCartney Rose

* *R. carolina* — Carolina Rose
  R. eglanteria — Sweetbrier
  R. foliolosa — White Prairie Rose
* *R. multiflora* — Multiflora Rose
  R. palustris — Swamp Rose
  R. setigera — Climbing Rose
  R. setigera var. *tomentosa* — Hairy Climbing Rose
  R. wichuraiana — Memorial Rose
* *Rubus argutus* — High-bush Blackberry
* *R. idaeus* — Red Raspberry
* *R. occidentalis* — Black Raspberry
* *R. trivialis* — Southern Dewberry
  (See Smith and Tucker for treatment of nine other species)
  Spiraea billiardii — Pink Spiraea
  S. tomentosa — Steeplebush

RUBIACEAE — Madder Family
* *Cephalanthus occidentalis* — Buttonbush

RUTACEAE — Rue Family
  Poncirus trifoliata — Trifoliate Orange
* *Ptelea trifoliata* — Wafer Ash
* *Zanthoxylum americanum* — Prickly Ash
* *A. clava-herculis* — Toothache Tree

SALICACEAE — Willow Family
* *Populus alba* — Silver Poplar
* *P. deltoides* — Cottonwood
  P. heterophylla — Swamp Cottonwood
* *P. nigra var italica* — Lombardy Poplar
* *Salix babylonica* — Weeping Willow
* *S. caroliniana* — Ward's Willow
  S. humilis — Prairie Willow
  S. interior — Sandbar Willow
* *S. matsudana* — Ball Willow
* *S. nigra* — Black Willow
  S. rigida
  S. sericea — Silky Willow

SAPINDACEAE — Soapberry Family
* *Koelreuteria paniculata* — Golden Rain Tree
* *Sapindus drummondii* — Soapberry

SAPOTACEAE — Sapodilla Family
* *Bumelia lanuginosa* — Gum Bumelia
  B. lycioides — Southern Buckthorn

SAXIFRAGACEAE — Saxifrage Family
* *Hydrangea arboscens* — Wild Hydrangea
* *Itea virginica* — Virginia Willow
* *Philadelphus hirsutus* — Mock Orange
  P. inordorus — Mock Orange
  P. pubescens — Mock Orange
  Ribes curvatum — Granite Gooseberry
* *R. cynosbati* — Prickly Gooseberry
* *R. Missouriense* — Missouri Gooseberry
  R. odoratum — Golden Currant

SCHISANDRACEAE — Climbing Magnolia Family
* *Schisandra glabra* — Climbing Magnolia

SCROPHULARIACEAE — Figwort Family
* *Paulownia tomentosa* — Princess Tree

SIMAROUBACEAE — Tree of Heaven Family
* *Ailanthus altissima* — Tree of Heaven

STAPHYLEACEAE — Bladdernut Family
* *Staphylea trifolia* — Bladdernut

STERCULIACEAE — Sterculia Family
* *Firmiana plantanifolia* — Chinese Parasol Tree

STYRACACEAE — Styrax Family
* *Halesia tetraptera* var. *monticola* — Carolina Silverbell

H. diptera — Snowdrop Tree
* *Styrax americana* — Storax
* *S. grandifolia* — Large-leaved Storax

SYMPLOCACEAE — Sweetleaf Family
* *Symplocos tinctoria* — Sweetleaf

TAMARICACEAE — Tamarisk Family
* *Tamarix chinensis* — Tamarisk

THEACEAE — Tea Family

Stewartia malachodendron — Silky Camelia

THYMELAEACEAE — Mezerum Family
* *Dirca palustris* — Leatherwood

TILIACEAE — Basswood Family
* *Tilia americana* — Basswood

ULMACEAE — Elm Family
* *Celtis laevigata* — Sugarberry

C. laevigata var texana — Dwarf Sugarberry

C. laevigata var *smallii*
* *C. occidentalis* — Hackberry

C. tenuifolia — Dwarf Hackberry
* *Planera aquatica* — Planer Tree
* *Ulmus alata* — Winged Elm
* *U. americana* — American Elm
* *U. crassifolia* — Cedar Elm
* *U. pumila* — Siberian Elm
* *U. rubra* — Red Elm

U. serotina — September Elm

U. thomasi — Rock Elm

VERBENACEAE — Vervain Family
* *Callicarpa americana* — French Mulberry

VITACEAE — Grape family
* *Ampelopsis arborea* — Pepper Vine
* *A. cordata* — False Grape

Cissus incisa — Marine Vine
* *Parthenocissus quinquefolia* — Virginia Creeper
* *P. tricuspidata* — Boston Ivy
* *Vitis aestivalis* — Summer Grape
* *V. cinerea* — Grayback Grape
* *V. palmata* — Red Grape
* *V. riparia* — Riverbank Grape
* *V. rotundifolia* — Muscadine

V. rupestris — Sand Grape

V. vulpina — Frost Grape

CLASS ANGIOSPERMAE
SUBCLASS MONOCOTYLEDONAE

LILIACEAE — Lily Family

Smilax auriculata
S. bona-nox — Catbrier
* *S. glauca* — Sawbrier
S. pumila — Sarsaparilla Vine
* *S. rotundifolia* — Greenbrier
S. smallii
* *S. tamnoides* — Bristly Greenbrier
S. walteri

INDEX

S

T

U

V

W

Y

Z